KEEPING DUCKS AND GEESE

KEEPING DUCKS
AND GEESE

Chris & Mike Ashton

A DAVID & CHARLES BOOK
Copyright © David & Charles Limited 2009

David & Charles is an F+W Publications Inc. company
4700 East Galbraith Road
Cincinnati, OH 45236

First published in the UK in 2009

Text copyright © Chris and Mike Ashton 2009
Photography and artwork © David & Charles 2009, except
those listed on page 126

A catalogue record for this book is available from the
British Library.

ISBN-13: 978-0-7153-3157-6 paperback
ISBN-10: 0-7153-3157-4 paperback

Printed in Singapore by KHL Printing Co Pte Ltd
for David & Charles
Brunel House   Newton Abbot   Devon

Commissioning Editor: Neil Baber
Editorial Manager: Emily Pitcher
Editor: Verity Muir
Project Editor: Caroline Taggart
Art Editor: Sarah Clark
Production Controller: Beverley Richardson

Visit our website at www.davidandcharles.co.uk

David & Charles books are available from all good
bookshops; alternatively you can contact our
Orderline on 0870 9908222 or write to us at FREEPOST
EX2 110, D&C Direct, Newton Abbot, TQ12 4ZZ (no stamp
required UK only); US customers call 800-289-0963 and
Canadian customers call 800-840-5220.

Contents

Introduction

Ducks and geese have been part of the rural economy for thousands of years, whether they have been farmed or hunted. Historical records give glimpses of how these birds have affected people's lives: providing warmth from their plumage, food from eggs and meat, and also cash from sales.

Although it was developed earlier and more extensively in the Far East, duck and goose farming has been documented most thoroughly in Europe and the USA. The special breeds have been conserved, giving a unique glimpse of farming traditions through the ages. About 40 breeds of domesticated waterfowl are presently recognized in the UK, and around 24 in the USA Standards, with numerous colour varieties. The rare breeds are popular with conservationists, pet keepers and show enthusiasts alike.

These are fascinating birds to keep and rear. Geese in particular have their own codes of behaviour and range of vocal expressions. They communicate much more effectively than most other forms of domestic poultry and show a keen awareness of body language, which is an essential component of their complex social existence. Tight family units and strong personal bonds make them ideal companions for human beings.

Keeping ducks and geese can be an immensely rewarding hobby. Children love these birds and, through them, learn how to relate to animals generally. Ducks and geese add a new dimension to a garden or smallholding. Whilst providing the potential for traditional produce, they show why they fascinated the famous biologist Konrad Lorenz, and were considered worthy of mention by the Greek poet Homer thousands of years ago.

A Brief History of Domesticated Waterfowl

Throughout human history only a small number of animals have been fully domesticated. In the birds this includes just four species of ducks and geese, all in the last 6,000 years. Eurasia provided the northern mallard, the greylag goose and the Chinese swan goose. The perching duck, *Cairina moschata* (the Muscovy) came from Central America. Of 140 species of waterfowl, these four formed the basis of all our farmyard breeds.

Human selection has meant that fully domesticated birds are quite different from their wild ancestors. They are frequently bigger, tamer, faster growing, a different colour or shape, and have longer breeding seasons. They have also been adapted to be more productive than they would be in the wild.

Humans originally took advantage of many different wild species, collecting the down of eider ducks from their wild nests in Iceland, and 'harvesting' pinkfoot geese by corralling them during their flightless stage. Greylag geese and wild ducks were depicted in Egyptian art, and the Romans wrote of both teal and mallards in their 'duckery'. There is little evidence that the mallard was fully domesticated in Europe until much later, although in the Far East distinctive duck breeds such as the Pekin and Indian Runner arose.

Domesticated geese

Early references to geese are difficult to find. Records of archaeological digs show greylag goose bones from Celtic times in Britain, but these are likely to be from wild birds. Mural decorations from Mesopotamia and Egypt are said to show the rearing of geese for ritual purposes as long ago as 2800 BC.

Roman accounts tell how the noise of the geese saved the Capitol from the Gauls. Pliny records flocks of geese being walked from France to Rome, and Lucius Columella recommends white geese, indicating that humans were selecting from the wild 2,000 years ago.

Geese were recorded on menus, in tenure payments and markets in medieval England, and Queen Elizabeth I dined on roast goose the day of the defeat of the Armada. But types of domesticated goose are not recorded in Britain

BELOW *Common Geese and Rouen ducks at a show in 1854. The Rouen from France was developed as a much larger, darker bird (Rouen foncé) in the UK. The birds illustrated are already quite large compared to the goose. They became even larger and acquired an oblong body shape by the 20th century. The geese are probably the breed found on the common land. The first National Poultry Show of 1845 classified geese as Common Geese, Asiatic or Knob Geese and 'any other variety'.*

until perhaps Gervase Markham in 1613:
'. . . the largest is the best, and the colour would be white or grey, all of one pair, for pyde [pied] are not so profitable, and blacks are worse.'

Geese were kept in the east of England in quite large numbers by the 1700s. In 1740, Lord Orford made a bet with the Duke of Queensbury that a drove of geese would beat an equal drove of turkeys in a race from Norwich to London. The geese did indeed win. As late as 1776, Pennant wrote that wild greylag goslings were being taken from the Fens, reared and made tame.

Ducks

From at least the thirteenth century wild ducks and geese were caught at the time of moulting. Nets were staked out and thousands of birds trapped in a single drive. Decoy traps on private estates were used until the late 1800s, when the drainage of the Fens and the development of the duck industry brought about their decline.

The great sea voyages, starting with Columbus, were the impetus for introducing new birds to Europe. The Muscovy arrived, and the Indian Runner and crested ducks probably came in Dutch ships from the Far East. Domesticated ducks feature in Dutch paintings from the seventeenth century and are the best indication of the way the mallard was being bred in Europe. In 1660, 'The Poultry Yard' by Jan Steen shows crested white and pied ducks, their type and colour indicating human selection. Another painting, by Melchior d'Hondecoeter, features crested Hook Bills.

In Britain ducks were not named as a breed; those on farms were known as the 'Common Duck'. The appearance of duck names and accounts in literature probably date from a reference to the Hook Bill in 1678; mentions of the 'Crook-Bill' and 'Normandy' (probably the Rouen) followed in 1750.

The rise of the domesticated duck industry in the 1700s was due to selective breeding of the 'English White'. Alison Ambrose quotes the Reverend St John Priest from 1810: 'Ducks form a material article at market from Aylesbury and places adjacent; they are white, and as it seems

of an early breed.' Together with the Rouen, Black East Indian and Call Duck, this white Aylesbury was among the four duck breeds to be standardized in Britain in 1865. The USA standard of 1874 included these four plus the Pekin and the Crested.

Exhibition

Exploration, the Empire, Victorian exhibition and a growing human population together formed the impetus for the development of specific breeds. Willughby described the Chinese goose in 1678; Albin illustrated the upright duck in the 1730s, and White Chinese were noted in George Washington's correspondence in 1788. In 1816 Moubray introduced news of the Embden to Britain. There followed the Sebastopol (reported in 1860), Toulouse, Hong Kong goose and Italian geese (1888); Black East Indian (1831), Indian Runner (1830s), Cayuga and the Pekin duck (1872 UK and 1873 USA). From this mix of exotic breeds new hybrids were developed and eventually became breeds in their own right. At the beginning of the 20th century, the demands of both utility and exhibition drove this expansion in a way that was probably unique to the period. The commercial hybrids of today depend upon that legacy.

ABOVE *The wild swan goose of Asia was domesticated in the Far East. The 'African' and Chinese geese show the same colour pattern.*

BELOW *This painting by J.W. Ludlow is from Lewis Wright's* Illustrated Book of Poultry, 1873. *The ducks, which belonged to Mary Seaman, were a prize-winning pair in 1870. Since then, the exhibition Aylesbury has been developed as a much larger breed.*

First Considerations

Waterfowl are not difficult to keep as long as you understand them. Take some time to find out about what they need to stay healthy. If you have the right equipment and allocate enough water and space for your chosen breed, the birds should have long and healthy lives. Ducks often live to ten years of age, and geese to twenty.

Understanding Waterfowl

All ducks, geese and swans belong to the family Anatidae. Within this family there are several 'tribes', which include the Anserini (swans and geese such as the greylag), Anatini (dabbling ducks, including the mallard) and the Cairinini or wood ducks (also called perching ducks, and including the Muscovy.) All have webbed feet, two layers of feathers and an extra layer of fat under the skin to keep them warm. These are all adaptations to their life on water, but the wild birds are also strong fliers and, in the case of the perching ducks, use trees as part of their environment.

Goose morphology

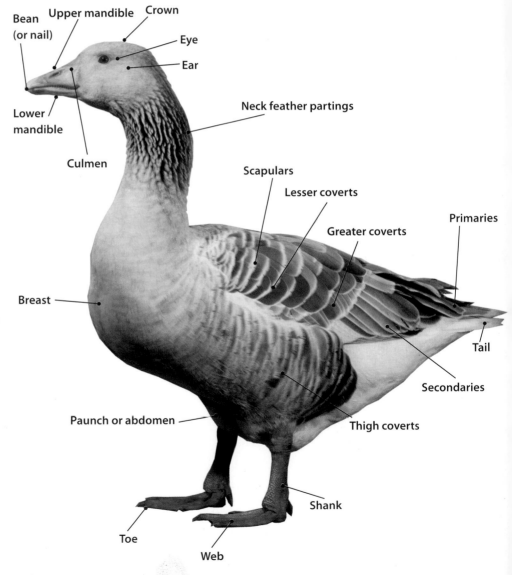

Bean (or nail)

Upper mandible

Crown

Eye

Ear

Lower mandible

Culmen

Neck feather partings

Scapulars

Lesser coverts

Greater coverts

Primaries

Breast

Secondaries

Tail

Paunch or abdomen

Thigh coverts

Shank

Toe

Web

Ducks have their legs set further back along their body than birds which live entirely on land. This feature becomes most exaggerated in diving ducks, whose feet propel them under water in search of food. Diving ducks and swans are quite ungainly on land and need deep swimming water.

Geese, on the other hand, mainly graze on land and are also quite happy walking. All waterfowl, however, tend to have their legs fairly widely spaced, for effective paddling. The set of the legs on the body means that all these birds, with the exception of the Indian Runner, tend to waddle rather than run.

Wild geese and ducks breed in the spring, and many species migrate in order to maximize safe places for nesting and food supply. Feather maintenance and replacement (moulting, see page 15) are timed to fit in with migration and breeding. The condition of the birds' feathers and stage of feather development affect their behaviour too. Feather type and condition are crucial to warmth, flight and migration, sexual display and nesting habits. Although domesticated birds no longer follow migration patterns, the annual cycle of reproduction and feather replacement is more or less the same as with their wild relatives, and very noticeable in domestic birds kept at close quarters.

Duck morphology

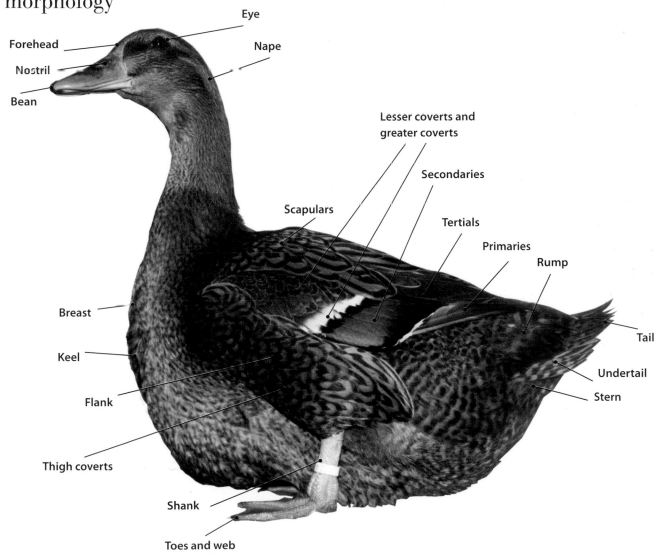

Feathers

We now know that feathers did not evolve first in true birds. Fossils from the Lower Cretaceous in China show that feathers appeared in non-avian theropods and so were inherited by the birds from the dinosaurs. Birds are also linked to dinosaurs by the fact that they lay eggs and have a similar eggshell microstructure. Other features once associated just with birds – such as down-like body covering, a broad plate-shaped sternum and substantial enlargement of the forebrain – are also now known to have arisen earlier.

Feathers are made of a protein-based material called keratin. This provides a structure very similar to that of scales, from which feathers may have evolved. As in hair and fingernails, the substance is 'dead', a product of once-living cells. It is hard, strong, light, flexible and warm. Additionally, it absorbs very little water, making it ideal for the many purposes and designs of feathers.

There are a number of different types of feather, even on a single bird, but they have major similarities in structure.

- Contour feathers cover much of the outer surface. They act as a 'shell suit' for waterproofing and insulation and give efficient streamlining.
- Flight feathers are longer and stiffer, and have little or no down. They are subdivided into the remiges, which comprise mainly the primaries and the secondaries on the wing, and the rectrices, which are at the tail.
- Down feathers have a greatly reduced or absent stem (rachis) and the 'velcro' mechanism on the barbs is more limited. Their main function is insulation.
- There are also bristles; semiplumes for insulation and bulking out the contour feathers; and filoplumes – fine, hair-like feathers with a tuft of barbs at the tip.

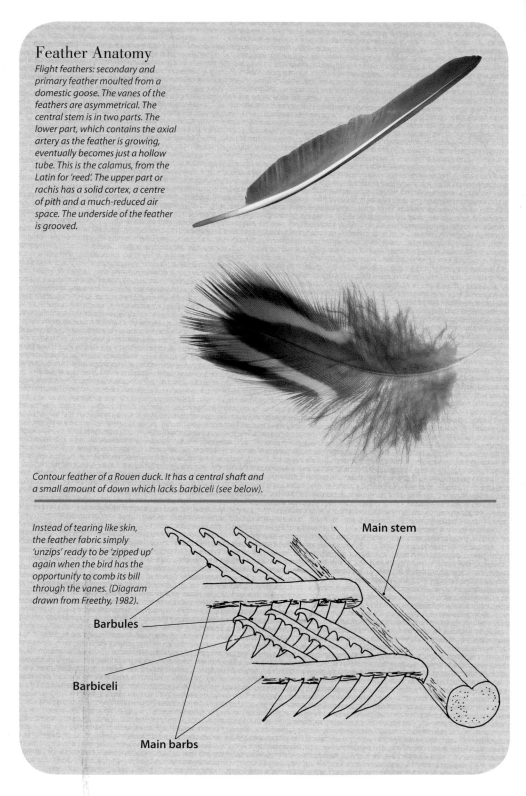

Feather Anatomy

Flight feathers: secondary and primary feather moulted from a domestic goose. The vanes of the feathers are asymmetrical. The central stem is in two parts. The lower part, which contains the axial artery as the feather is growing, eventually becomes just a hollow tube. This is the calamus, from the Latin for 'reed'. The upper part or rachis has a solid cortex, a centre of pith and a much-reduced air space. The underside of the feather is grooved.

Contour feather of a Rouen duck. It has a central shaft and a small amount of down which lacks barbiceli (see below).

Instead of tearing like skin, the feather fabric simply 'unzips' ready to be 'zipped up' again when the bird has the opportunity to comb its bill through the vanes. (Diagram drawn from Freethy, 1982).

Main stem

Barbules

Barbiceli

Main barbs

Replacing feathers: moulting

Over the course of a year feathers get fairly 'tatty'. They are weakened by exposure to the elements and altercations with other animals. So the outer 'shell suit' and flight feathers have a complete moult. The downy feathers are retained for insulation, though they are plucked from the breast and underparts by females to line their nest.

Most ducks and geese have one moulting pattern in common: they shed their flight feathers on the wings in one go. This differs from the so-called 'sequential moulters' who lose just one or two flight feathers at a time, so that they retain the power of flight. Loss of flight would be lethal for raptors: they would starve. In contrast, flightless geese can still feed and look after their young (also flightless at this stage), and they can escape predation by taking to the water.

Geese keep moulting simple; males and females all do it together, or around the same time. They have a single type of plumage, and moult once during the year. Mallards (and domesticated ducks) moult over a more extended period, and at different times.

When geese moult

A mother goose drops her flight feathers when the goslings are just 2–4 weeks old. The regrowth of these feathers is so important that all other feathers, including the wing coverts, wait until the flights are regrown. Wild geese renew their flights in 21–24 days, so by the time the goslings are 6–8 weeks old, the flight feathers of adults and young are at a similar stage of development.

The female goose takes responsibility for incubating the eggs, though the gander hangs about while she is sitting. After the goslings hatch both parents play a role in protecting and rearing them. Pair bonds are very strong, meaning that there is no need for fancy plumage to attract a mate, so geese moult their body feathers only once a year.

Ducks are different from geese

Not only do ducks moult twice a year, but the males and females moult at different times. Females are in camouflage plumage all the year, but drakes must change their appearance depending on seasonal demand. Smart feathers are needed for display in the breeding season, but camouflage is better for the summer, especially when the males have no flights.

Although wild mallard will pair-bond for the breeding season, the sexes generally split up at sitting and rearing times. In the breeding season, bright, sexy plumage is an advantage for the male; a fit drake in glossy plumage will attract a mate. The green sheen on his head, the claret breast and the shining speculum between the black and white wing bars all signify a healthy bird. Females will definitely choose mates like this which advertise their fitness to breed.

Marginal coverts

Primaries

Primary and secondary underwing coverts

Secondaries

Tertials

Axillars

ABOVE *Brecon Buff gander showing off his underwing feathers. He has ten flight feathers (primaries) which are the most asymmetrical, a set of secondary feathers, and tertial feathers which cover and protect the important flights when the bird is at rest. Axillar feathers fill the air gap in flight. Smaller underwing coverts cover up the base of the larger feathers, ending with tough little marginal coverts which are very firmly attached.*

Mating behaviour

In wild populations, female mallards choose the most colourful partners, using inciting displays to demonstrate which drake they want. In domestic ducks, females use a series of loud, single-note 'gak' calls to indicate they are looking for a mate. Unattached young females cruise around on the water and adopt a behaviour pattern known as 'nod-swimming': they swim forward, neck stretched out, then sharply bring it back. This display, with the neck lowered, is similar to the mating pose, and is a stimulus for the males to display as well.

It's at this point that the drakes show off their fine feathers. They burp, whistle and do head-up/tail-up displays. A sudden lifting of the head, wings and tail stretches the shiny feathers of the neck and flashes the speculum at the interested female. In domesticated birds, this behaviour is most marked in Calls, but larger breeds do it too.

Unfortunately, if there are a number of drakes, such displays lead to several males raping ducks, which can be badly mauled and even drowned. Domestic females must therefore be protected from the attentions of too many drakes.

Drakes show an obvious change of feathers at the close of the breeding season. While the ducks sit and incubate, the drakes go into eclipse. Their flashy winter nuptial plumage is moulted and replaced by 'duck feathers', giving them camouflage before they moult their flights. This is the opposite way around to geese.

In contrast to geese, wild mallard drakes take no interest in family rearing, and can even be a nuisance for ducklings. Drakes often congregate in 'boys only' groups while the females sit.

Males have a very obvious second moult in autumn, probably caused by temperature changes. The eclipse plumage of the body feathers is moulted again, and the bright autumn nuptial plumage resumed.

And what about the female? Like a sitting goose, she retains her flights and delays moulting. There is no point in growing new feathers while her food reserves are being used up in laying, then incubating, eggs. Only after the ducklings are well grown does she drop her flights and replace her body plumage.

ABOVE *White domesticated Mallards in Australia. The male grasps the female's head feathers in order to balance during copulation. She has been helping him by spreading her wings.*

BELOW *Two Runner drakes (right, with green bills) in July. They already have new flight feathers and eclipse plumage which mimics that of the females. The females continued to lay and did not drop their flights until the end of August. Female Call ducks finish laying earlier and moult their body feathers and flights by midsummer.*

Bathing, cleaning and preening

Ducks and geese have at least one cleaning session a day. They dive forwards into the water, submerging their head and flinging water over their back. They roll sideways, even overturn, and beat their wings on the water's surface. This keeps the important flight feathers in good condition and also creates a cooling shower on a hot day. Keepers of pet waterfowl all know how ducks love a hose-pipe shower in hot weather, sipping the water particles in the air and then going in a mad rush around the pool. Ducklings introduced to a pool of water for the first time will duck and dive, rushing round and under this new medium that they should enjoy for life. That's why intensive duck farming is so difficult and incompatible with welfare; it really cannot provide the duck's natural environment.

With the bathing routine to remove the dirt comes the beginning of the preening process. This is the complex business of reconditioning the feathers; it involves thoroughly inspecting, cleaning, lubricating and arranging all the feathers at least once a day. Preening removes parasites and scales which cover newly sprouting feathers, and spreads oil from the preen gland (see next page) over clean feathers. It also involves combing the barbs on a feather so that the small hooks called barbules and even smaller barbiceli all interlock to form a protective layer to repel water. Waterfowl do this combing by vigorously stripping their flight feathers through the bill, starting at the quill and working outwards along the barbs that form the vane.

ABOVE *A Roman goose conditioning her feathers and removing parasites such as scale lice. These lice cling to the feathers and eat scales of skin. A few lice on the bird do not cause damage, but their numbers need to be kept under control.*

LEFT *Hook Bill ducks beating their wings on the water as part of the cleaning process.*

Using the preen gland

For waterfowl especially, secure waterproofing is vital. Ducks and geese need to keep their feathers in top condition to protect themselves against waterlogging whilst swimming, and in wet weather. They do this by using an oil-secreting gland found on the rump near the base of the tail.

The preen gland exudes a thick liquid of fatty acids, waxes and fats, the composition of which varies according to the time of year. Stimulation by the duck's bill encourages this liquid to flow so that oil is transferred to the lower mandible and the head region.

The bird strokes its lower bill over the breast and underparts, then rubs and rolls its head over the preen gland and the body feathers to spread the oil evenly. This allows water to run off in droplets – like the proverbial 'water off a duck's back'.

It may be that the oil of the preen gland has little to do with directly 'waterproofing' feathers, its main function being to keep the feathers supple, though it is also believed that the oil helps kill bacteria and fungi. However, one only has to look at ducklings to see that preen-gland oil does have some function in repelling water. Ducklings reared under a mother duck have natural waterproofing from contact with her feathers.

Feather structure too is important in repelling water. Constant attention to zipping the barbiceli together helps to form a waterproof shield. It has also been suggested that the mechanical action of preening applies an electrostatic charge to the oiled feathers, which may repel the water.

Wet feather

Whatever the truth about preen-gland oil – that it repels water itself, or that it only creates the conditions for maintenance of an effective shell-suit – ducks even with a preen gland can get out of condition and suffer from a problem called 'wet feather'. When things go wrong, a bird looks distinctly damp instead of glossy. The feathers become waterlogged and unable to repel water. The outer contour and flight feathers are affected. In a severe case, even the under-down may start to suffer. The bird becomes cold and miserable, and avoids water, which can then make the condition even worse.

The commonest causes of wet feather are dirt and mud. Once the feathers have become coated with soil particles or muck, the bird rapidly loses condition. A wet environment exacerbates the problem. Affected birds simply cannot dry out, and cannot recondition their feathers by preening. The solution is to bathe the birds in really clean water, then give them the opportunity to dry out and recondition their feathers. In really bad cases, only the growth of new feathers next season cures the problem.

Occasionally, if a single bird is affected, it may be that the preen gland is not producing enough oil. This may be caused by a poor diet: if you feed cheap or out-of-date food, the bird may not be getting enough vitamins. Adding green food (grass clippings) and whole wheat can help, as can ensuring that the bird's environment is clean and removing any parasites. If birds are on genuine free range, they will pick up their own animal protein and greens and should stay in good condition. It will probably not matter what they are fed as a supplement to their main wild diet.

Vegetable and mineral oils will not help birds suffering from wet feather, though pure Purcellin oil, used to recondition oiled, rescued seabirds, may.

BELOW *The position of the preen gland is shown by the tuft of oily, yellow feathers at the base of the tail feathers, on the bird's rump.*

Planning and Commitment: what to consider

Looking after livestock is a responsibility that should not be undertaken lightly. If you have no previous experience of keeping ducks and geese, it is important to find out about the behaviour of the different species and breeds. Do make sure that you have enough space and a suitable environment for the right type of bird. Appropriate facilities will give you security and peace of mind.

How many birds?

Geese and ducks are social animals that are accustomed to living as pairs in the breeding season or, for much of the year, in a flock. Although many households keep only one dog or cat, birds benefit from a partner and it is unfair to keep a bird on its own. Adult birds are often sold in pairs (a male and a female) for this very reason.

It might seem a rather attractive idea to have a single goose or duck as a pet. A duckling or gosling reared on its own will become imprinted on its owner but the human partner cannot always be around and the bird will be distressed in the human's absence. So, pet birds should always be reared in pairs or groups. Spending time with them will keep them tame anyway, and they will be much happier with their partner for company: they can follow their normal behaviour with a companion of their own kind. However, it is advisable not get too many to start with, and certainly not too many males.

Space

Space is a prime consideration: waterfowl need more space than poultry, unless you make a considerable investment in all-weather surfaces and ponds which can be readily refilled. Geese do need to graze, yet ducks need space for a different reason: if they are confined on a small patch, they can quickly damage the grass in wet weather by probing in the soil for worms. Two average-size ducks might be all right on a patch of garden about 10m (11yd) square, but this depends very much on the type of soil and the availability of water.

Whilst a pair of geese does not need half a hectare (an acre), they do become overweight if kept in confined conditions on food solely from bags. A pair of small geese needs a minimum grassy area of 10 x 20m (33 x 66ft) if they are housed at night (to keep the grass cleaner) and they also have access to a pond or plastic pool.

For more details about space and numbers, see page 26.

RIGHT *Give geese plenty of space, especially in winter, to encourage good grass growth.*

Security

Birds need security from predators and from traffic. Firstly, they must not have access to vehicles. Call ducks are so blasé they may easily be run over. A gander may attack moving car wheels, with minimum damage to the car but maximum damage to himself. Ducks and geese live on the ground (or on water) and do not perch in trees, so they must also have a secure shed for overnight accommodation. A large pond or lake, even if it has an island, is not secure: foxes may catch birds on the banks. It will take some time for the birds to learn that safety is to be found on the water and by then many may be lost. In addition, mink are aquatic and will take birds.

Also invest time in planning fencing; a secure area is needed to accommodate birds when they first arrive. They must be prevented from running away from the unfamiliar place.

Choose a suitable height and mesh: dog- and fox-proof mesh will need to be higher than fencing designed just to keep birds in.

Foxes can leap a fence 1m (3ft3in) high.

Circumstances in your area – how much land you have, water supply, frequency of predators etc. – will determine how much freedom the birds can have to free-range, or if they need to be confined.

Routine

Your daily routine will vary according to the seasons. Birds will be shut up at night for much longer hours in winter and in severe climates, where you also need to ensure that the drinking water does not freeze. Checking the birds at least twice a day is not time-consuming, but it is important to be around to release them at a safe time in the morning and to shut them up before dusk. If the timing and amount of daylight make this difficult, then you need a vermin-proof enclosure. Unlike poultry, which put themselves to bed at dusk, ducks and geese are quite happy to stay outside. Train them to return to their shed each night at a regular feeding time, drive them inside and close the shed up. Timing devices that are used to close poultry sheds do not work with waterfowl: the birds are likely to end up being shut out and left to the mercy of nocturnal predators.

Check food consumption regularly. Don't put out too much at once, only what the birds will eat over half or a whole day. Feeding time is also the best time to observe the birds and their behaviour: you will notice failure to eat, difficulty in walking and other signs of illness early on.

Water supply can be from buckets and bowls, as long as you refill these on a daily basis. A pond can be a moveable container, or permanent, as long as it has a sump for emptying and cleaning. Best of all is a small stream, but that can have its attendant security problems.

Collect eggs every day in the laying season, and put fresh litter in the nesting place. As long as topping up the litter keeps night quarters dry, you need only clean the shed once a week. Deep litter, cleared out just once or twice a year,

RIGHT *This situation looks idyllic, but these birds must be housed at night.*

can be used as long as the bedding is odour-free and dry.

Going away

Stock needs looking after 365 days a year, and this will undoubtedly affect your leisure time unless you have someone who can help out. Keeping birds is a big commitment.

That's one of the reasons why it's important to start off in a small way. Having just a few birds, and the routines and predator-proof set-up described on the previous pages, makes it easier for someone else to cope while you are away, and it also gives peace of mind.

Costs

Setting up can be costly, though the price will vary depending on whether you employ a contractor, and also the type of materials. The enclosures need not fill all the land available, but they must be large enough to keep the birds safe and well whilst you are away, or during short days in winter.

Further expense will be incurred in finding food and water containers and in making all-weather surfaces if only a small space is available. Ponds, pond construction and water pumps all cost a great deal.

If you are planning to incubate eggs and rear ducklings in any number, you will need further equipment such as heat lamps, brooder areas, rearing coops and protected netted areas. These are likely to cost more than you can recoup. Running costs for feed, water and so on are almost certain to outweigh anything you can earn from the production of eggs or table birds because it is very difficult to beat the costs of large-scale enterprises. Pure breeds of geese and ducks on a small scale are there to be enjoyed for what they are: as part of a smallholding with high welfare standards; or as show birds or pets.

Remember that bird keeping on a small scale is a hobby. Welfare is foremost and this is now enshrined in legislation in the Animal Welfare Bill in the UK. In 2009, keepers of hobby and companion animals and birds will have 'Good Practice Guidelines' which take into account a suitable environment and diet for the birds, and also their quality of life. Similar legislation exists in Europe but is patchy in the US. Books such as this one provide basic advice, as do various bird societies. If you get really keen on the birds, it is worth joining a duck and goose organization such as the British Waterfowl Association to see what they have to offer, to meet like-minded waterfowl keepers and to benefit from their information and experience.

Neighbours and noise

Waterfowl are not suitable for keeping in a confined space in close proximity to neighbours. Quite often there are regulations on keeping birds on housing estates. Do find out if restrictions apply before you spend money on birds or equipment. Neighbours may also object to noise, which in certain breeds can be considerable. Whilst the quacking of Call ducks may be music to their owner's ear, the neighbours may disagree. So buy a quieter breed if there is potential conflict. Unlike cockerels, drakes are the quiet sex, and a couple of those may be more desirable as pets than females. Chinese geese are very vocal, and Africans are high up the decibel scale in comparison with their European cousins. Find out about the breed characteristics before you commit to buying.

Some people keep birds on allotments and this can work well if several of you have birds and can co-operate with release and shutting up. At certain times of the year, ducks can be released to eat the slugs, but will need to be confined when crops could be damaged. There is always the risk of vandalism to fencing and damage to or theft of the birds. Consider the circumstances and security of your allotment first.

LEFT *Call ducks are more vocal than other breeds. If quacking females are likely to be a problem with neighbours, just keep drakes or choose a quieter breed.*

Getting Started

Ducks and geese are hardy creatures; they need
housing mainly to protect them from predators.
They are tough enough to stay outside all year
in the West European climate but need help
if there is a big freeze in a continental winter,
or a prolonged heat wave. They are creatures
of habit and can be trained (or they train you!)
to come back to feed in the same place each
evening. But introducing birds, especially
ducks, to a new place needs care in planning
their facilities.

Environment

Space and numbers

Ducks have traditionally been reared in areas with gravelly river bottoms and sandy soils. There is a good reason for this: ducks can make a mess on waterlogged clay where their beaks probe the soft soil for worms very easily. Geese will improve the grass: ducks can destroy it and promote weeds. So start in a small way or with small birds to see how the land responds, and consider ways of managing the land. Too many ducks will result in a muddy lawn in winter unless there is alternative wet-weather accommodation, and they will also eat all the garden produce (and the slugs!) unless they can be moved onto different patches of land according to the season.

If you are planning to keep ducks and you only have a small space, invest in an all-weather gravel pen. In dry weather and in the summer when the ground is not moist for very long and the grass repairs itself quickly there is no problem. However, in a wet spell in winter, ducks are best confined to a small gravel, deep bark-covered or concrete area, or stabled to prevent them doing their worst.

For geese, the successful allocation of land will depend on the weather, the season and the type of soil. Free-draining sandy soils are the best for geese in order to minimize diseases and to keep a good turf. Ideally geese should be kept in an orchard or on a smallholding where they can graze in rotation with sheep and cattle. [Note that geese and ducks should not be kept in confined conditions with larger animals: they can get trampled.] If the grass is not short, it will have to be mown, for geese only get good nutrition from short, sweet grass. They will also eat garden produce and chew saplings if these are not protected.

BELOW Grass should be kept short for maximum nutrition. Do watch out for hazards: rivers can rise and flood, and you should check that any old farm equipment doesn't pose a risk of injury or entrapment.

Sex ratio: keep the ducks happy, don't get too many drakes

Introducing new chickens to a flock of hens can be difficult, because the 'pecking order' has to be observed. Adult ducks are different. A new female who is put down in a group of females is generally well accepted after a short period of inspection and greeting, and ducks are unlikely to inflict serious damage on each other. Occasionally, in male-only groups, drakes may persistently scrap with one another and have to be separated.

The exception to good behaviour in ducks is where there are too many drakes. Females should never be kept with a lot of drakes. Two or more males which repeatedly mate the same female are a nuisance and a danger, and need separating, even if this means that surplus males are put in a 'drakes-only' pen. A new female needs to be watched if there is more than one drake in the flock to make sure that she is treated well. Remove the drakes if they are a problem and leave only one until the new duck is established

Mating is rather inelegantly but appropriately called 'treading'. This is a precarious activity without the active consent of the duck. Unfortunately, with too many drakes, the ducks literally get run down and bedraggled, as they try to escape the attentions of the drakes. Ducks can die in these conditions; they may suffer from damaged eyes and prolapse of the oviduct. Too much mating does not do the drakes any good either; they can suffer from prolapse of the penis. So it is best to avoid problems by not keeping too many drakes in the same area. Although it may take more time and work, birds should be kept in small groups to safeguard the females.

However, if you want to keep drakes as quiet pets or slug foragers, they will probably get on amicably as long as they have been brought up together and there are no females around for them to compete over. Introduce them at the same time so that they are on equal terms. A large group of males may fight and cause damage to each other during the breeding season, so may need be separated into smaller groups, but plenty of space and foraging helps kept the peace. If you are keeping them just as pets, it's probably best to have no more than two.

Geese are less of a problem. They are normally sold in pairs and like living in pairs. However, in mixed sex flocks a higher number of females is advisable. Ganders will fight in the breeding season, but the top gander usually establishes order if there is enough space. Birds which do not get on, which can be both male and female, must not be housed together in confined conditions.

Security fencing

New birds may need to have their wings clipped (see page 37) and be safely fenced in, otherwise they can fly or wander off. Don't allow them access to a river or stream until they have got their new home and night quarters firmly established in their minds, otherwise they might simply get lost.

Fencing in Call and miniature ducks is also very useful during the breeding season, when a female may go broody on a well-concealed nest and not return home. Call females are frequently lost through this mismanagement.

A 1m (3ft) fence will keep in all but the most determined or frightened ducks, though many breeders use 1.8m (6ft) fences with electrical protection top and bottom to keep foxes out. This type of fencing really needs to be installed by a contractor.

The size of the mesh you require depends on the size of bird. Pig netting will restrain only the very largest birds; most duck breeds need rabbit or poultry netting.

The best permanent fencing is 12 gauge/5cm (2in) weldmesh. It's twice the price of poultry netting, but requires minimal post support and will last for years. Backed by two or three strands of independent electric wire and sunk into the ground, it provides protection from dogs and badgers. With electric fencing supported on independent posts on the outer side, it will also be reasonably fox-proof. Its disadvantages are that ducks can put their head out through the mesh

ABOVE *This 1m (3ft 3in) proof fencing is well protected by two electric wires at the base, and a strand at the top. Vegetation should be kept clear of the wires.*

LEFT *The Hook Bill ducks and Chinese geese are protected by 1.8m (6ft) high predator-proof fencing.*

and mink and small polecats can squeeze in, so if you live in an area where these predators are a risk, go for 3.5cm (1½ in) poultry mesh instead. Be aware, too, that geese can get tangled in temporary electric mesh fencing and may die.

Protection from predators

It is essential that birds be securely housed at night if there is no fox-proof and badger-proof fencing. Waterfowl do not put themselves to bed but, unlike chickens, they can be herded to their hut at any time.

Many people believe that birds will be fine at night on the pond. This *can* be so when the ducks and geese spend the night safe on an island or on the water and are accustomed to these conditions. But when new birds are introduced, losses will occur until the survivors have learned that their refuge is on the water. It takes time and significant losses for the birds to become street-wise. More females will be lost than males; they do not move as fast when in lay, and they may also nest on the ground in unprotected positions. Domesticated birds are also exposed to duck viral enteritis, potentially carried by the wild mallard. Buzzards and hawks may swoop down on small and young birds; rats, herons, magpies and crows will also take ducklings and goslings.

ABOVE LEFT *A netted area, such as a fruit cage, is very useful for protecting Call ducks or small young birds.*

ABOVE *Wild birds will take advantage of free food. This includes pheasants and more aggressive birds such as crows. Use the containers described on page 41 to reduce food waste and contamination.*

Housing and bedding

Poultry housing is now made in some beautiful and sophisticated designs, with raised nesting areas, perches and walkways. This is suitable for chickens but not for ducks and geese because they do not perch and rarely use raised nesting boxes. All that waterfowl need is a secure wooden shed with a good ventilation panel which faces away from prevailing winds and rain. So as well as checking poultry housing suppliers for sheds, visit garden centres and agricultural stores to see what is available. Garden sheds and dog sheds are often more suitable and cheaper than specialist poultry housing.

Many sheds are made of tongue and groove boarding treated with a preservative. Modern wood preservatives are supposed to be safer than old-style creosote but if preservative has been applied recently, make sure the fumes have gone before housing birds. Tanalised wooden sheds from a reputable supplier are the best.

There needs to be enough floor space for the birds to be comfortable at night. A dog house 1.4m (4ft 6in) square will accommodate up to ten Call or four large ducks, or a pair of geese. Ducks and geese should not be housed together.

The house should be positioned in a corner so the birds have to run into it, not round it. It will last longer if raised on a dry base. The door should be fairly wide so that ducks do not run over each other; poultry pop holes are not suitable for driving in ducks or geese. The doorway needs a ramp if it is not nearly level with the ground.

Site the shed on slightly raised, dry ground. Avoid the floodplain of a small brook. It's surprising how many duck keepers have lost birds and equipment in flash floods.

For ventilation use a weldmesh panel high up on the sheltered side of the house. This is better than at ground level where the birds can be bothered by vermin. Make sure that the mesh will not admit polecats or mink. Shut birds up each night before dark, otherwise they will almost certainly fall prey to foxes.

Bedding can be any soft litter such as wood shavings or roughly cut, coarse sawdust. Do not use hay or dried grass clippings as these encourage the growth of fungal spores and aspergillosis. Straw is likely to encourage a build-up of ammonia; white-wood waste is better. Make sure that it is from untreated wood i.e. it does not contain pesticide/or fungicide. White-wood shavings for horse bedding are readily available. If the birds are not fed and watered in the shed, the bedding will stay drier and, in the winter in particular, you can top it up rather than changing it so that it builds up to a thick, warm layer.

The used bedding can be composted; however, it takes a long time for wood shavings to break down. After a year on the compost heap they can be used for top-dressing to deter weed development. Although hemp bedding composts more readily than shavings, it is not advised for waterfowl. If eaten, it swells up to many times its original size. Never use it with young birds.

BELOW *Housing must be fit for purpose. This coop and polythene-covered run houses young goslings (or ducklings) to protect them from bad weather and predators (see also page 109).*

ABOVE *This shed is placed in the corner of the pen so that the ducks can be driven in easily at night. The security fencing is 1.3m (4ft) high poultry mesh, very well secured to posts and rails.*

Water

Ducks and geese must have good access to water (see pages 17–19). At the very least, they should be able to wash their face and eyes and throw water over themselves to wash and also condition their plumage by using their preen gland. Diving and wing beating are also part of the cleaning routine. Without sufficient bathing water ducks can become dirty and suffer from 'wet feather' when the feathers fail to repel water (see page 19).

Natural water is very inviting, but not necessarily best. A small stream is better than a river; rivers in flood can sweep birds away and can also harbour mink. Small streams without fish in them (fish attract mink) are safer. Ducks keep natural ponds clear of weed, and the birds will find a lot of their own high-protein food. However, ponds with decaying vegetation can become a source of botulism in a hot summer.

Small numbers of birds can manage quite happily with just buckets and bowls, though they must have clean water at least once a day. Washing-up bowls are all right for Calls, as long as they can get into them. However, Calls appreciate water so much that they stay in much better condition on a pool.

Ponds can be made by sinking a polythene liner into a hole in the ground, but these soon become fouled by droppings. It is better to buy a strong paddling pool or children's sandpit and empty and refill it on a regular basis. It can also be relocated on clean ground each time it is replenished, reducing the risk of disease.

Always make sure that birds can get out of water containers. Smooth-sided ponds can be hazardous for ducks. If the pond is full, the birds just walk or flap out, but if the water level is low, young birds cannot get out. Use a floating wooden ramp for steep-sided pools. Always fence young children out of the duck area in case of accidents.

ABOVE *Ducks and geese need water to drink and bathe, however it can also be deadly for them. Always ensure that the birds can get in and out of ponds safely.*

tip

Birds can drown in narrow, steep-sided water containers. Containers should be wide enough for birds to turn around in, and so not get trapped.

BELOW *Low, wide buckets such as this one are cheap water containers, ideal for ducklings and goslings as well as adults.*

Where and When to Buy

Pure breeds are normally sold in pairs (a male and female). It is much easier to start with a pair of adult birds, or at least a pair over 16 weeks old. Very young ducklings or goslings will need protection and well-thought-out rearing facilities. Larger adult birds are more 'weather-proof' and less vulnerable to predators.

Birds in full plumage are generally available from August onwards. They can be bought at specialist sales and auctions, where they are not always graded for quality.

It is often better to buy direct from the breeders, many of whom advertise in specialist magazines (see page 126). They also belong to organizations with websites, some with breeders' pages. These websites have become increasingly popular and give access to producers around the UK and the USA. Breeders associated with bona fide organizations are more likely to have good examples of the breeds than other sources because they exhibit their birds and know the correct breed characteristics. Their birds may be more expensive than some of those at auctions, but these breeders will generally give good advice and be more reliable about the breed, age and sex of the birds.

Local newspapers and feed-stores are also useful sources of adverts and advice. Just be careful: not all advice will be sound and not all birds will be pure.

Specific breeds may be hard to find. Some pure breeds are rare, locally and globally. Hook Bill ducks, for example, were rescued from extinction in Holland in the 1980s, and are still rare internationally.

Nearly all pure breeds of geese are rare. They used to be kept on many smallholdings and farms, but increasing urbanization and commercial farming have reduced the numbers. They still remain in areas where there are small farms and, quite often, they are associated with hobby breeders who show their birds. Be prepared to order birds in spring and to travel some distance to get them. With rare breeds, expect to pay as much for young as for adults; they are always in short supply. Birds bred in greater numbers are cheaper at summer and autumn sales and there is more

LEFT *Young Silver Appleyard drakes moulting their first feathers at ten weeks of age. Their plumage quality cannot be assessed until they are 18 weeks old. The greenish tinge to the bill, the white collar just appearing and the green feathers sprouting on the head show that they are male.*

ABOVE *Colour faults may only show in fully feathered birds. This cross-breed buff goose has a white breast, and the grey and whites may be females with sex-linked colour. Only breeders who keep records can guarantee such a breed (see pages 64–65)*

choice. By the spring, all stock will be limited. Also, breeders will have incurred additional costs by overwintering their birds, and prices will consequently be higher. Only the best birds are generally kept on to this stage, often as reserve breeding stock. If you want extra females, order them early in the year.

Commercial birds

Large numbers of females only are available in commercial duck breeds, such as Khaki Campbells and commercial whites. The whites are often termed 'Aylesbury' but they are in fact carefully bred hybrid ducks designed to lay lots of big white eggs. There are also white table strains which are larger. None of these is the Aylesbury. Commercial ducks are bred in their thousands and in the egg-laying strains the males are culled at a day old, which is why females are available on their own. So, for high-production egg-layers, source these birds by looking in journals covering commercial stock (see page 126). The ducklings are generally available between March and July.

Commercial geese can also be found in this way, though some of the major producers rarely seem to advertise.

Commercial birds are sometimes sold as POL (point of lay). In ducks, this means that they are coming up to 24 weeks of age. The term is irrelevant to geese which mainly lay in the spring.

THINGS TO CONSIDER

- **Good examples of pure breeds are more difficult to obtain than pet-quality birds.** Do not rely on hatching eggs, because they may not produce what you expect. Begin by purchasing the correct adults, and then think about hatching later.
- **The waterfowl breeding season is shorter than that for poultry.** Geese, on the whole, lay in spring to early summer (February to June). Ducks have a longer laying season but their peak breeding time is April – the same as the wild mallard. The season can be extended in commercial birds, which lay more eggs annually, but they reduce hatching production in autumn and winter. Plan to buy your stock according to the season and its availability. Young fully feathered adult birds, over 16 weeks of age, will not generally be available until the summer.
- **Visit an autumn waterfowl show to see the birds, find out what they're like and talk to other owners.** There may be pens of birds for sale. Rare Breeds centres sometimes keep ducks and geese, but these are often limited in number and quality compared with what breeders can exhibit at a major show.
- **Find out if there are any suppliers near you,** by looking through magazines, local adverts and the internet.

ABOVE *Rearing goslings needs special care, but if you do it well the birds should stay tame for life.*

Ducklings and goslings

Producers of pure breeds are less likely to sell very young stock than commercial producers. That is especially so with keepers of top-quality birds who wish to grow them to select their show birds and future breeders. They may be prepared to sell stock which does not meet their requirements, but this does not often include young ducklings and goslings, except in easy-to-hatch breeds.

If you do buy ducklings or goslings, pay special attention to rearing conditions to give them a good start in life and to avoid problems such as slipped wing and weak legs. Use the right food at different stages of growth, and protect them from parasites, bad weather and predators. If young birds have already been reared outdoors they may carry parasites and it is essential to be aware of this hazard and know how to deal with it. The advantage of young ducklings and goslings, of course, is that they can be hand-reared and stay tame. Do not buy just one on its own. See pages 106-109 for more details.

A pair of adult birds

Buying an adult pair usually allows you to get the sexes right. Buying ducklings and goslings on impulse, or incubating eggs, can mean you end up with too many males which will need new homes or to be culled. Whatever you do, do *not* dump males on the local lake or river. It is irresponsible and cruel.

RIGHT *Even at 4–5 weeks, the concentrically fringed rump feathers of this Silver Call show that it is a drake.*

> **tip**
>
> Don't buy more drakes than ducks and be prepared to keep adult birds in groups with no more than one or two drakes, depending on their behaviour in the breeding season.

Telling the sex in ducks

- Sexing adult ducks is easy: the ducks give a loud 'quack' and the drakes make a characteristic scraping noise.
- In ducklings, the 'peep' becomes a female quack around 5–6 weeks of age.
- Drakes also have the characteristic sex-curl feathers on the rump from about 14 weeks, though these can be removed by unscrupulous dealers or just by wear and tear.
- Apart from black and white breeds, there are also sex differences in coloured feathers. Most drakes have a characteristic 'hood' of green, blue or glossy brown.
- Another key area is the rump, where coloured breed males often show darker marking. In males with patterned feathers on the rump, e.g. Abacots and Silver Calls, the feathers are fringed concentrically with a paler colour, whereas the females develop a central streak along the length of the feather.
- In many breeds drakes have green bills, and females have orange, brown-orange or dark bills. In white breeds the bill is often orange-yellow in both sexes, but the beaks of females become darker with age.
- Pictures from the Waterfowl Standards will show typical colours of the breeds and sexes and aid identification; these will also indicate the expected size and shape of the birds.
- Birds can be vent-sexed (examining the sex organs at the anus/vent) if necessary. They should only be examined by experienced handlers.

Telling the sex in geese

Sexing geese is not as easy as sexing ducks, especially with the larger breeds such as Toulouse. Geese have to be sexed by size and behaviour. Close observation by owners will establish which are males and females: the geese always know!

- Experienced breeders will vent-sex geese at around 3–5 weeks of age and close-ring their pure breeds.
- Male goslings tend to grow faster, be more confident in coming forwards and have larger feet, which shows up when slipping closed identity rings over their feet at around four weeks.

- When picked up, tame females will chew clothes more than males do.
- From around 12 weeks of age, the females' voices are deeper; ganders' voices remain higher. Female Chinese develop a characteristic 'oink'. Ganders usually end up larger and females have a heavier undercarriage.
- White breeds such as Embden, Roman and Sebastopol are often auto-sexing as infants (that is, they are born different colours, so you can tell them apart at a glance). Males and females can show a faint grey-back pattern in the fluff at hatch. This fluff is darker in the females, which can look like Grey Backs at this stage. These markings are replaced by white feathers, but females may retain some grey feathers on the rump until their second year. White Chinese goslings are all bright yellow: their white gene is different from that of European geese.
- Auto-sexing breeds such as West of England and Shetland also show the grey-back (pied) pattern in the females and retain it as adults. Pilgrim geese likewise hatch grey females, which have a darker beak than the males. Colour is a reliable discriminator only in pure bred birds.
- Some breeds show no difference in colour at any stage, e.g. Brecon Buff, Steinbacher and Toulouse.

Quite often, the sex of young geese seems to remain unknown until the breeding season. Females then lay eggs. These are laid on a cycle of around 36 hours. So an egg a day from two birds denotes two females. This does happen!

Two ganders which have got on well together until spring may, at this point, begin to fight. In the absence of a female, one will try to mate with the other and neither will submit. This should not be confused with a goose demanding attention by 'roughing up' her gander until he mates with her.

In a larger group of birds, surplus males will be driven out by dominant ganders. Unlike drakes, ganders will generally stick to their partners. They do not engage in indiscriminate rape. Also unlike drakes, they will develop a strong family bond with the goose and goslings, and are altogether much 'nicer' characters.

BELOW *The sale of downy ducklings and goslings at markets and auctions is not good practice. Most sales will not accept birds unless they are well feathered. Keepers have a 'Duty of Care' to provide good conditions – warmth, food, and water – at all times. These conditions are not met when ducklings are sold in this way at such a young age. Buy baby birds directly from the producer, or with the foster mother. They will also be healthier if they have not been chilled and dehydrated for long periods during transport and marketing.*

Checking the birds

A further advantage of buying adult birds is that you can assess their size and shape. It is at this stage that deformities of the spine, for example, become apparent, though spinal deformities of the neck and wry tail may appear earlier. Make sure that the birds move normally and do not limp, and handle them to check they are not too thin.

Catching should be done by quietly driving a group into a shed and then cornering the bird. Also check the birds' eyes and feet. Make sure the eye is clear with no opaque growth, and that the undersides of the feet are not damaged. Check the vent, too: a dirty vent is a sign of illness.

BELOW *A deformity such as this arched back (seen on the bird in the centre) may only show up when birds are almost fully grown.*

LEFT *How to hold a bird: slip a hand under the body to lift the bird's weight and hold its legs; at the same time tuck it under your arm. Don't squeeze the legs together, and never lift waterfowl by their legs.*

Clip up to six flights below the coverts

The new quills are filled with blood at 4–9 weeks and so cannot be clipped.

Mature feathers can be clipped on one wing: just the five outer flights, before the coverts which overlie them.

CLIPPING

- **Clipping wing feathers to deter flight** should be done when handling birds at point of sale, or when they first arrive at their new home. The feathers must be fully formed. Feathers grow from quills filled with blood which develop between at 4–9 weeks. These feathers must not be damaged until growth is complete.
- **Clip only the outer 4–6 primary feathers of one wing** (see the feather diagram on page 14). This unbalances the bird so that flight is limited. Clip the feathers only up to the wing coverts. This will avoid cutting through the hollow part of the quill and leaving it open to dirt and bacteria. If in doubt, cautiously clip only the first 5cm (2in). Clipping does not hurt a bird; it is similar to cutting hair. It is not the same as pinioning, which is not done in domesticated birds.
- **Only birds which fly readily need clipping.** This includes Bantam and Call ducks and the light breeds of geese such as Roman and Czech. Sebastopols cannot fly, Chinese don't seem to attempt it and the other breeds are usually too heavy.

Transport and boxes

Transport regulations concerning licences do not apply to hobby birds. However, welfare regulations do. Birds must be transported in suitable boxes which are well ventilated and which will allow them to stand up (see photograph right).

Cardboard boxes can be used if you cut several air holes in them, but are not suitable for transport in a hot car without air conditioning. Adapt boxes by using a wire-mesh top held in place by electrical tags. Cardboard boxes must not be stacked; they can collapse. Don't put boxes in the boot where air circulation in low. Birds quickly overheat in cars in hot weather, and may die.

The best bedding for transport is clean straw; it is less dusty than shavings and gives a better grip for the feet.

Getting them home and settling them down

When the new adult birds arrive home, put them straight in their new shed to settle down after the journey. Although they should not be fed and watered in their shed on a regular basis, it is a good idea to give them a bowl or bucket of water with a handful of wheat in it if they have travelled on a hot day.

Make sure that ducks especially are securely penned in a wired run to start with. If they have come from a place with a lot of birds, their first instinct will be to find some. If they are unused to other animals they will also be upset by dogs, however good the dogs may be; the birds need to get used to the situation gradually.

Once they have had a chance to settle in, find out what food they like best. They may have already acquired a taste for bread, so feeding them a favourite titbit will help to tame them. The best way to tame birds is to let them get a bit hungry. Do not leave food out ad lib; teach them to associate food with your approach and feed it in small amounts to start with.

ABOVE *Commercial plastic-mesh crates, dog cages and pet carriers are suitable for transporting, especially in hot weather. Park your car in the shade when loading. Never leave birds in a stationary car in hot weather.*

Feeding Waterfowl

Digestive system

Ducks and geese are different genera and do not have exactly the same feeding habits. However, they do have a similar digestive tract. It differs from that of the chicken, which has a sac-like 'crop' to store food in the oesophagus before it enters the gizzard. The crop is useful as a food store for overnight digestion. Waterfowl simply have a widening of the lower oesophagus into the proventriculus where the food is softened by digestive juices

The digestive tract next enters the body cavity (through the wish bone) and reaches the gizzard. Mammals do not have this muscular stomach; they have teeth to macerate their food. Modern birds do not have teeth, instead their 'chewing mechanism' is located within the gizzard itself. This massive muscle inside the body cavity uses grit and sand in place of teeth to break down food physically. Angular grit such as granite grit is much more effective than well-rounded particles. The gizzard muscle itself is lined with a tough material which protects it from wear and tear.

Ducks, and especially geese, require the grit to break down hard grains of cereal and to macerate vegetation. This is particularly important where grass forms a large proportion of the diet. Geese have to work very hard for a living because they rely on vegetation as their main food source and they lack a crop (though the gullet and proventriculus will store some food). They also lack fermentation processes which break down plant cellulose as in the rumen of cattle. Instead, grit and sand in the gizzard have to puncture grass and release cell sap, the main source of nutrition.

The food can pass backwards and forwards between the proventriculus and the gizzard, and preparation is also aided by digestive juices. Once the food has been prepared in this way, it passes through the intestine and nutrients from the food are absorbed into the blood stream.

Geese do have particularly large caeca. These are two blind-ending branches of the lower intestine; their function is not fully understood. The caeca contain bacteria which may aid the break down of plant cellulose.

Eventually, the waste food is expelled thought the vent as parcels of chewed grass and caecal waste, which is darker and less frequent. Geese are inefficient feeders, requiring a greater pro rata intake of food than ruminants. They have a very fast throughput of food, measured at around two hours. That is why they are very useful as lawnmowers, especially in awkward places where machinery cannot reach.

BELOW *Ducks have a smaller body weight than geese, and also eat concentrated nutrients in the form of invertebrates as well as grass. Geese spend more time grazing, and it essential that their gizzards are kept in very good order to digest the large intake of vegetation. Make sure they have sand and grit. They should be wormed regularly against gizzard worm (page 114).*

What do wild ducks eat?

Ducks are omnivores and also generalists. They will eat whatever is available: land and aquatic flora and fauna. Mallards typically feed mostly on plant food, including seeds and grass, but for reproduction, growth and moulting they need the high nutrient levels provided by invertebrate food. Reproduction and the growth cycle are timed to coincide with the best available food. Females can eat twice as much as males, and also consume more macro-invertebrates during their egg-laying period. Gastropods and crustaceans probably supply a large part of the minerals needed for egg formation.

The dabbling ducks, including Mallard, typically feed at the water's edge and in less than 20cm (8in) of water, where they find emerging mayfly, caddis and midges. They up-end and also dabble and puddle in shallow water, mud and wet soil to find tadpoles, snails, worms and grubs.

ABOVE LEFT *A duck's mouth is designed to find food in liquid. The bill contains corpuscles which are extremely sensitive to vibration, making it easy to detect live food. Water and mud is sieved through the lamellae (strainers) at the side of the bill, enabling the duck to find out what is there.*

ABOVE RIGHT *Adjusting the position of the upper and lower bill during feeding alters the size of the gap between the lamellae and allows the duck to determine the size of the food particles drawn in.*

ABOVE *The muscular tongue functions as a suction tube and also has sensitive fringes which can sort out food from detritus. The tongue and beak are able to vibrate at a rapid rate during the foraging process: just listen to domestic ducks puddling in the mud on a wet day!*

Food for domesticated ducks: pellets and wheat

Domesticated birds like to behave in the same way as the wild mallard, but are usually kept on insufficient land to provide enough natural food for their needs. Also, unlike wild ducks which may lay only one or two clutches of eggs a year, domesticated ducks can lay over 200. Their diet has to be adjusted to take account of this high nutrient output, so feed them cereal grains and pellets. This is a very unnatural diet compared to the wild mallard, but has been carefully tailored to suit their needs.

Pellets are a compound food manufactured from cereals (mainly wheat, but also barley and maize), vegetable oils and soya for additional protein. Vitamins, minerals and, in the best quality products, fish meal (which is particularly good for growing birds) are added. The milled materials are bound together and the result is dehydrated to stop it going mouldy. Pellets are dry food, so if you feed your birds on them, ensure that they have free access to water.

Dry mash for hens is unsuitable for ducks because it clogs up their nostrils and mouth. Feed it only if it has been mixed up to a stiff porridge first; it must then be eaten up at once, as it quickly goes mouldy.

In addition to pellets, give adult ducks whole wheat as 50 per cent of the diet; it is higher in nutrients and protein (up to 12 per cent) than maize (corn), which makes both ducks and geese too fat. Given the choice, ducks will eat more pellets, and drakes more wheat.

Diets for breeders and layers

Pellets are manufactured in different types depending on the age and condition of the bird. Layer and breeder pellets contain more protein (16–17 per cent), calcium (2–4 per cent) and phosphorus than grower pellets. Breeder pellets are better quality than layers, and contain more vitamins. Very young birds need starter crumb rations which are higher in protein, and then move on to grower pellets (see page 106).

Laying ducks must be fed a ration of layer pellets which contain calcium and phosphorus in the correct proportion for eggshell formation. Insufficient minerals result in the birds taking the minerals from their own bones, or laying soft-shelled eggs which have a membrane but no shell. Genuinely free-range birds may still find enough invertebrate life to fulfil their nutrient intake but they too should be offered layer pellets morning and evening. Drakes do not need high-calcium pellets.

Ducks for breeding need a better diet. Just as the wild mallard females seek a high-nutrient diet in spring, the breeder ducks should receive a special diet for a strong embryo. Breeder pellets have no undesirable additives (unlike some hen layer pellets which contain artificial colouring) and more vitamins.

In practice, many people use ordinary poultry layer pellets and wheat all year round for adult birds but vary the ratio according to the season. In cold weather, ducks tend to consume more wheat.

ABOVE *Allow ducks to rummage in pasture for grubs, worms and slugs.*

All ducks benefit from finding some of their own food in the vegetable patch, garden, pond, stream and pasture. They also like additional chopped greens (grass, cauliflower trimmings and chickweed) in a bowl. Lettuce is not a good food, as some varieties can be high in nitrates.

Quantities

Do watch what the birds want to eat and vary the type and amount of food accordingly. Big ducks eat a lot, up to 240g (8oz) a day in large, growing breeds. The average is more like 120g (4oz), but females in lay need more (especially pellets) than non-productive drakes. Call ducks have tiny appetites, but even they need more in cold, wet or windy weather. Birds housed at night eat less than those which live out in a fox-proof pen.

If adult ducks have access to wheat under water all day, pellets can be fed economically. Give them only as much as they will clear up in 20 minutes each morning and evening rather than leave this more expensive food out, especially in wet weather.

Food containers for ducks and geese

Pellets (or wheat and pellets mixed) should be fed dry in a heavy bowl, e.g. an old cast-iron casserole or trough. Call ducks need to be fed in heavy, shallow containers; buckets are more suitable for Indian Runners and geese. Pellets should *not* be fed in water: they disintegrate. Damp pellets also go mouldy, and the mould is a hazard, so the container should be protected from the weather. If the birds are smart, the food container can be kept inside the shed away from wild birds – don't leave food with the birds overnight. Don't leave water inside the shed either: the bedding becomes wet, and becomes a health hazard.

Ducks and geese can also have access to wheat fed under water outdoors. This will prevent sparrows, pheasants and pigeons from taking the food. Ducks and geese like to feed under water, and this also keeps their eyes clean.

Food storage

Manufactured food has a sell-by date, generally less than two months. This should be stamped on the bag, on the label or both. The label will also give the ingredients and the purpose of the food. Always check the use-by date on a bag of pellets, because vitamins deteriorate with age. Buy food from a store with a good turnover and clean, dry, cool storage conditions. At home, store food in a cool dry place to prevent mould developing. Operate a food rotation system, using the oldest first. Keep the bags on a pallet if necessary and invest in vermin-proof food bins. It also pays to clear up spillage and keep some rat or mouse poison in the storage area in a container accessible only to the vermin.

THE IMPORTANCE OF GRIT

- **Ducks should always have access to mixed poultry grit** especially if they have some wheat in their diet. The insoluble grit (flint) helps to break down food in the gizzard, and the chips of limestone and seashells provide additional calcium. Grit is obtainable from feed merchants and country stores. Do not use oyster shell alone; it does not provide essential insoluble grit.
- **If in doubt about availability of minerals** for birds, lime the ground with hydrated lime or calcified seaweed, available from garden and country stores.
- **Geese also like mixed poultry grit** in their egg-laying season, and should be provided with coarse building (concreting) sand all year round. The muscular action of the gizzard then punctures the grass, crushing the cells and releasing cell sap so that they are both subject to further digestion.

ABOVE *Geese enjoy grasses such as ryegrass and timothy when kept short. The serrated edge of the beak is adapted to cut vegetation efficiently.*

Food for geese

Swans and geese are herbivores. Wild geese spend most of their time searching for marine and saltmarsh vegetation, grasses and sedges and even roots (tubers and rhizomes), berries and lichen. The choice of food varies with the species, time of year and location in the annual migration pattern. Favoured plants under cultivation are rye grass, timothy and white clover; and wild geese in both Britain and the USA have adapted to feeding on spilled grains and the shoots of cultivated grains. In general, a mobile wild population seeks out shoots which are short and growing and therefore high in protein. Protein demand is highest during egg production; the protein requirement of geese

breeding in captivity has been assessed at 18 per cent.

Wild geese also obtain their nutrients for breeding from vegetable matter rather than crustaceans and insects. A proportion of the minerals required for egg formation is built up in the bone before migration, but most is accumulated in medullary bone (which acts as a reserve for calcium exchange for eggshell formation) when the birds reach their breeding grounds. Fat reserves built up prior to the breeding season are progressively used up during egg-laying and incubation.

Unlike ducks, domesticated geese are often not keen on eating pellets for extra nutrients, even in the breeding season, unless this is a

habitual part of their diet. Domesticated geese generally lay fewer eggs than domesticated ducks. If a goose is allowed to follow her natural instincts in spring and lay one clutch of around 12 eggs and then go broody, good grazing will supply her nutrient demands. She should also have access to coarse sand and mixed poultry grit. This is important for the geese as their nutrition comes from the grinding and puncturing of grass with the little stones in the gizzard when the juices are released for digestion. Geese will also eat other vegetable material such as potatoes and carrots if they are accustomed to them.

However, some geese will lay 40 eggs and not go broody, and some commercial strains

have been selected to lay even more. Greater egg production does result from feeding supplementary duck/goose breeder pellets containing protein. Supplementary food should also include whole wheat which can be fed under water (in a bucket).

How much to feed

Feeding geese too much can lead to a build-up of fat in the body cavity. The intestine becomes narrow, its folds separated by too much fat. Overfeeding also leads to liver damage and difficulty in laying. Geese are healthier and live longer if grass forms part or most of their diet. Check their weight frequently to see if they are too fat or too thin and need better feeding or worming. There should be breast muscle on either side of the breast bone, which should not stick out sharply.

Growing geese on zero grazing will consume up to 450g (1lb) of food per day; adult geese around 225g (8oz) a day. Geese on quality grass may need less that 120g (4oz) per day.

Grass quality is important. It must be short, with new shoots. These have the highest food value in the spring and early summer; in the winter more supplementary food is needed. Clean ground is essential.

Although geese, and especially goslings, eat copious amounts of grass, the pasture needs to be managed. In drier areas in the USA and Australia, this includes seasonal irrigation. Grass also stays cleaner if the geese can be rotated around two or three patches, and housed at night. This limits the amount of droppings on the ground, and the manure can then be composted.

Goose pasture should also be mown for weed control; they will not eat docks, thistles or plantain. Protect garden plants, as geese have an appetite for garden produce, and they will also chew and destroy saplings. Birds should not have access to poisonous plants such as foxglove, nightshade, ragwort and laburnum, though they will probably not touch daffodils and laurel.

Food suppliers

A local feed supplier can be found by looking in the Yellow Pages under Animal Feeds. More of the large food manufacturers now produce duck/goose rations and it is worth checking with the main office for Smallholder Feeds, BOCM-Paul's Marsden's Range and Marriages to see if the right kind of food is available in your area. This is especially important if you are rearing ducks; some coccidiostats added to poultry grower rations are not good for ducklings and can kill Call ducklings.

ABOVE LEFT *Geese need some supplementary food as well as grass. This can be wheat, but offer them breeder-quality pellets mixed with wheat in the breeding season. Put the mixture dry into a heavy container.*

ABOVE *A goose's beak is deeper and shorter than a duck's. It is designed for snipping and pulling leaves, shoots and roots, and has serrated edges. The lamellae have become cutters rather than filters. The tongue also has spikes, inclined backwards, rather than the sensitive fringes of the duck which are designed to find invertebrates rather than cut and swallow large amounts of vegetation.*

Choosing the Right Breed

PURE BREEDS OR HYBRIDS?

- **Ducks and geese have been kept for centuries** for their utility qualities, and distinctive regional types, with their own individual temperments, have evolved around the world. Some of these have become established as pure breeds (and colour varieties). Often they are quite rare, and they survive mainly because enthusiasts work at preserving rare breeds.
- **Pure breeds cost more to buy and rear** than commercial birds, which are cross-breeds, often referred to as hybrids. Pure breeds are usually kept in small numbers in a more natural environment on smallholdings and in large gardens. The birds have more space, live longer and eat more food. Unlike hybrid birds, which have 'hybrid vigour', pure breeds can become inbred unless breeders take considerable time and effort in sourcing new stock, sometimes from abroad.
- **Despite the greater cost of pure breeds**, and their less efficient food conversion rates, they are often preferred for small flocks and as pets, where food production is a minor consideration. The birds are more colourful and distinctive and have an added value: they or their offspring can be exhibited or sold as pure breeds. They are much more fun for the family and it is well worth spending some time on finding out about their characteristics and personalities before deciding what to buy. For more details on the individual breeds, see pages 50–65 for geese and 68–89 for ducks.

Pure breeds of ducks

Large ducks such as the Aylesbury and Rouen were originally 'table birds'. They are now a heavy exhibition type, much slower growing than commercial cross-breeds, and real plodders. They are very tame if hand-reared. In this respect, they are very easy to manage but they eat a lot of food and definitely benefit from a pond.

Other large breeds of ducks invented later in the 20th century are not as enormous and are better layers. They include colourful breeds such as the Appleyard and Saxony. These move faster and lay more eggs than Rouen, and are also useful table birds, especially in a good strain of Appleyard.

Lightweight ducks, including the Indian Runner, are the best layers, but move fastest of all. Commercial Khaki Campbells can clock up over 300 eggs per year. Slightly larger 'exhibition' strains of Campbell do not lay quite as well but, like their cousins the Welsh Harlequin and Abacot Ranger, they are expected to lay over 200 eggs per year and can reach 280. These figures apply to young birds which are healthy and not too inbred. The advantage of commercial Campbells is that they come from flocks with a wide gene pool to preserve their 'vigour', with a new dash of Fawn Runner added from time to time.

Runners are popular, but vary a lot in their capabilities. Trout runners generally do not lay as well as a good strain of White or Blue Runner, but they live longer. It is definitely the strain that counts: carefully chosen crosses of two unrelated strains of Runner can produce a very good laying bird; inbred types are poor. Runners can manage with less water than most breeds of ducks. They are very active and are happy free ranging on pasture, but if they have a pool they will use it. They also need careful handling as ducklings to keep them really tame.

If you want ducks mainly as pets, the very big or very little ones are your best bet. Bantam and Call ducks especially suit smaller gardens very well. The Call is 'the little bird with the big voice': only invest in these if the neighbours won't mind the noise. Bantams are usually much quieter. They are also good sitters and betters layers than Calls. Both are good fliers and need clipping on one wing when they are moved. They may even need clipping at a permanent home. Black East Indians and Miniature Appleyards are both very beautiful and should be readily available if ordered from breeders.

Muscovies are a different species from the other breeds of ducks. They are large, especially the males, good fliers and very strong. The smaller females are good sitters and very quiet, and the drakes also can only give a breathy hiss. Large drakes are not compatible with other breeds; they are far too heavy to mate with other ducks, and will try. So keep Muscovy males separate from other breeds, or just keep the females.

Pure breeds of geese

Whilst ducks may need protection from the attention of children, young children may need protection from adult geese. Unlike ducks, geese can be strongly bonded and very family-minded, and so can respond to a perceived threat. This may include the body language of young children: their impetuous movements make the geese nervous. The birds also know they can take advantage of the child's smaller size. Some children have a sympathetic interest in birds and animals at a very young age, but in most cases geese are best avoided until children are at least seven years old and are able to gauge how other things behave.

Goslings are a delight to rear; however, they are very fragile when young and if children are looking after them they will need supervision. Contact with humans makes the birds very tame indeed and they will follow their keeper just as wild goose families follow their leader. Most hand-reared geese stay tame for life but occasionally ganders can change their character. This is more likely in commercial white geese which can become the archetypal 'farmyard gander with attitude'. Africans, Steinbachers, Pilgrims and some strains of Brecon Buffs generally remain much more amenable.

Commercial white goslings are available only for a short period between February and May. They cost around four times as much as commercial ducklings and the price rapidly rises as the goslings grow.

Pure-breed goslings are also hatched in spring, but it may be difficult to prise them away from their owners. They are often kept to keep the goose happy, to eat the spring and summer grass, and simply because the producers like them. They may only be sold from August onwards, when they are fully feathered and can be assessed for quality. You may have to order pure breeds in advance, and even wait a year for the best quality.

Geese have a far shorter laying season than ducks and produce fewer eggs which, in pure breeds, are more difficult to hatch than other waterfowl. The lighter breeds such as Chinese and Roman can produce up to 60 eggs a year; a more typical figure for other breeds is 30. Apart from commercial white geese, geese are kept mainly as hobby birds for exhibition purposes, as pets and for lawn mowing.

LEFT *Exhibition Brecon Buff geese can be hard to find.*

Exhibitions

If pure breeds are your choice, do visit an exhibition to see them. Birds are sometimes available there in sale pens. These exhibitions are listed in magazines and at www.waterfowl.org.uk. See page 126 for information on how to find out about exhibitions near you. Most of the major shows are held during the autumn and winter months, when the birds are in full adult plumage and can be properly assessed by qualified judges who keep and breed waterfowl themselves. Each recognized breed in the Waterfowl Standards will have its own breed class at the biggest shows, which range over the country. Germany, Australia and the USA have their own waterfowl standards, but, for long-established breeds, the standards are almost the same the world over.

Show birds must be of the correct colour for their breed, so a specialist show will introduce you to good examples. They should all be clean and in good condition, and these days they are often marked with closed numbered identity rings on their legs. This is useful for keeping breeding records. These rings are permanent, but no other rings are allowed on the show pen birds.

Shows can be a great opportunity to assess stock, meet breeders and birds, to get a copy of the Waterfowl Standards and to join a waterfowl organization.

BELOW RIGHT *Tall Embdden ganders at the European Waterfowl Show in Holland, 2002.*

BELOW LEFT *A 'Best of Colour' Fawn-and-White Indian Runner drake at a Championship Show. Larger shows can have between 400 and 1000 ducks and geese, and up to 120 Indian Runners.*

Geese

These very intelligent birds have never been commercially exploited in such numbers as chickens and ducks. They behave quite differently from other farm birds, retaining a strong family bond which extends to the humans who raise them. Geese have good memories of each other and humans, as well as of situations and places. This is why, if well-handled, they will stay tame for life, and will repay you with their gardening and their talkative and inquisitive nature. Darwin was wrong when he said, 'No one makes a pet of the goose.' Konrad Lorenz got it right: '...geese exhibit a family existence that is analogous in many significant ways to human family life.'

Keeping Domesticated Geese

Geese have been domesticated for thousands of years and are found in a variety of climates. There are suggestions from both Roman and Celtic cultures that the goose was regarded as not only useful, but sometimes sacred, and also as a companion pet. Geese are vigilant, intelligent and sociable birds, which are sadly underestimated and misunderstood by many.

Productivity

Geese are not as productive as chickens and ducks, but are very adaptable, given an outdoor life. They mostly lay between mid-February and early June, producing about 30 eggs a year, though Chinese geese and their commercial crossbreeds may lay for longer, producing up to 80 eggs. The eggs are laid on alternate days, often in two clutches, because a goose, given the right environment, will normally go broody after her first clutch of 12–18.

A goose egg needs around 30 days to hatch, but 32 days is not unusual, and the goose must be looked after very carefully while she sits. Having tame birds is a great advantage at this time; and only tame birds should be allowed to rear goslings. In the absence of such parents, goslings should be hand-reared to keep them tame. They are then easier to manage as both farm birds and pets.

Domestication has increased the weight of the birds so much that only the lighter breeds are really suitable for natural incubation; the larger breeds can squash and kill their goslings at hatching, so it's better to remove the eggs to be hatched in an incubator (see page 98).

Geese take longer to mature and breed, and longer to grow, than chickens and ducks: they are not fully mature until two years old. For these reasons they are expensive to produce, rear and buy as table and breeder birds. On the other hand, if they are kept in small numbers and fed largely on grass, they can be both useful and cheap to maintain.

Learning to live together

Many geese are now kept as pure breeds and pets largely because of their behaviour and personalities. They are intelligent, social animals with a keen awareness of body language. They relate to one another in complex ways and have a range of vocal and physical signals. They know each other individually and they recognize people even when transported to new locations. They also have very good memories. If they have plenty of considerate human contact as goslings, they will remain tame for the rest of their life. Animal training is largely about training humans.

Putting your head near to a goose or gander is a signal of trust and friendship. It is what members of the family or flock do to one another, and it is a useful tip for potential goose owners. Approach quietly and allow the bird

LEFT *Different breeds get on well together, but cross-breeds will result if eggs are incubated from mixed groups.*

to get close to you. Unless you are very sure of its temperament, do not put your hand out to stroke it. Your hand may be perceived as a threat (it is the same shape and posture as the outstretched neck of an aggressive bird) and geese really do not enjoy being stroked as cats and dogs do. They much prefer to do the 'stroking', in this case by chewing shoe-laces, trousers, coats, jewellery and beards. They become very quiet when their head is near yours, scary as that may seem.

All ganders get hormonal at breeding time, especially if they are guarding a laying or sitting goose. If they threaten or attack, the worst thing you can do is get aggressive back. Being macho with ganders just does not work. If anything, it increases their adrenaline or testosterone levels and drives them into a greater fury. The best thing to do is use lateral thinking and a little psychology. Showing parental affection can really deflate the aggression. Picking up an aggressive gander (carefully and firmly), cuddling it for a few minutes like a gosling and then putting it down quietly can have a surprisingly calming effect. Don't try this until you are very confident of picking up geese and always do it with tame birds first. Failing this, just keep out of the 'territory' and do not 'tease' hormonal birds. Immature males are worst in this sort of situation.

A word on gosling behaviour: goslings hate to be on their own. They love company, especially that of their parents, their siblings, indeed any suitable living thing. They just hate solitude. Please avoid keeping a single young bird. It will certainly imprint on its owner and get terribly lonely whenever the 'foster parent' has to go to work; later it may become quite neurotic when it comes to socializing with others of its own kind. Keep two or more together and see how they get on.

TEMPERAMENT AND BEHAVIOUR

- Most of the breeds detailed in this chapter belong to the same species, and the differences between them are largely in colour and size. Differences in temperament or behaviour are more likely to be the result of the strain used to produce the birds, or the way they were handled when young, than any inherent characteristic of the breed itself. Birds bought from someone who rears a thousand birds will behave differently from birds who have been reared in a group of six or ten, whatever the breed.
- For this reason, it is important always to buy from a reputable breeder – see page 32 for advice on how to find one – to see how the birds behave before you choose them. And do please read the chapter on Getting Started carefully before you make any decisions.
- Any breed-specific characteristics that may make a goose unsuitable for beginners are listed in the individual entries and on page 46.

Keeping birds tame is part of the scene. I know it is tempting to let nature take its course, but there are many dangers attached to letting geese or ducks hatch and rear their own young. We try to hatch our birds under reliable broodies and finish off in designated incubator hatchers. The ducklings and goslings spend their first few days in boxes under heat-lamps in our kitchen. They get used to humans walking, talking and handling them. This way we can monitor any medical or developmental problems. We lose far fewer babies than we would if we let the geese trample the tiny goslings or the ducklings fall prey to the crows and magpies. Additionally, the babies become tame and develop into tame adults.

Whatever you choose, do not let nervous or 'wild' adults bring up their young. They will only generate even wilder offspring due to a combination of genes, upbringing and role models. We find that proven foster parents are beneficial for young geese. Ganders in particular that come up to be fed, enjoy the company of humans and allow themselves to be handled inculcate the same sort of behaviour in their young charges.

RIGHT *A pair of Sebastopol geese showing the characteristic curled feathers.*

ASIATIC BREEDS

AFRICAN AND CHINESE GEESE

fact file	
Goose breed	African
Classification	Heavy
Weight range (male)	10–12.7kg (22–28lb)
Weight range (female)	8.2–10.9kg (18–24lb)
Colour	Grey wild-colour
Wild ancestor	Swan goose
Origin	China
Status	Watch (US)
Eggs	Up to 30

fact file	
Goose breed	Chinese
Classification	Light
Weight range (male)	4.5–5.4kg (10–12lb)
Weight range (female)	3.6–4.5kg (8–10lb)
Colour	Grey wild-colour
Wild ancestor	Swan goose
Origin	China
Status	Watch (US)
Eggs	Pure breeds 30–40, utility Chinese up to 80

ABOVE *Young Chinese are active, alert birds. Ganders (the two at the front) are taller than females and develop a larger knob on the head.*

Both of these domestic breeds are derived from the wild swan goose (*Anser cygnoides*) (see page 9), which has breeding grounds in eastern Russia, Mongolia and China. Nearly all the rest of the world's domestic breeds are developed from the European greylag (*Anser anser*). The two wild species are thought to be closely related, having evolved from a common ancestor about 360–400,000 years ago. Although not likely to interbreed much in their natural habitats, they can and do crossbreed in captivity. These crosses are the foundation of many Russian breeds of domestic goose.

The wild swan goose has a similar plumage colour to the domestic forms, yet its body shape is almost horizontal, less upright. Its bill is proportionately longer and it does not have the fleshy lump in front of the crown.

In the Far East there are dozens of local breeds of domesticated swan geese, whilst in Europe and America the extremes of size are represented by just two breeds, the Chinese and the African, the Chinese being small, slim and elegant while the African is massive. The African does not originate in Africa but is really just a large form of the Chinese. It has a huge frame and a smooth dewlap below its chin. Both breeds have velvet-smooth necks

and swan-like 'knobs' at the front of their skulls, larger and more obvious on the males. Males are heavier and quieter than females, which make a characteristic 'oink' call and are quite noisy. This can be important if you have sensitive neighbours, although some whisky distilleries employ gangs of Chinese geese to warn of intruders.

Both breeds are extremely confident and become excellent pets. They graze effectively and enjoy tugging at roots, though they seem to be less destructive of young saplings than other breeds. It still pays to put bark protectors on valuable slim trees whenever geese are around! Other advantages of keeping the Asian geese include their lack of inclination to fly: it is possible that they have lost the ability altogether. Body and wing proportions (like those of Indian Runner ducks) might make this difficult. Commercially, Chinese geese have a reputation for a small,

ABOVE *A pair of exhibition African geese.*

compact carcass, handy for the table. The young females can also be extremely good layers: although most pure goose breeds lay up to about 30–40 eggs in a season, utility Chinese, which are bigger and less graceful, can lay 80 eggs per year, some of those in the autumn. Their eggs are easier to hatch than those of exhibition Africans, which are very large: over 210g (7½oz).

There are white versions of Chinese and African geese, although the White African is now rare in Britain. Other interesting plumage colours include the Buff African and the Blue varieties of both breeds.

The normal wild-colour is called 'grey' or 'brown', very similar to the wild swan goose. Both of the Asian breeds have a distinctive brown stripe that covers the top third of the head and runs down the whole of the back of the neck. The front of the neck is light fawn fading to cream; the breast and underbody are also light fawn, fading to white at the stern. The wings and shoulder feathers are similar to both species of wild goose: a darkish grey-brown edged with a much lighter colour, again almost white. The bill, knob and eye rings are black, although, if they have been crossed with White or even Buff varieties, they may show elements of orange, which is the basic skin colour in those forms.

Africans lay fewer eggs, perhaps up to 30, the number depending on the strain and the age of the bird. There are 'super Africans' and just ordinary Africans and there is a big difference between large exhibition and utility strains. The utility strains are lighter in weight and are likely to have been crossed with Chinese.

BELOW *White Chinese are very elegant on water. They were recorded in the USA in 1788 and in the UK in 1848: '…of a spotless pure white   more Swan like than the brown variety, with a bright orange-coloured bill, and a large orange-coloured knob at its base'.*

WHITE GEESE

For those who don't know about ducks and geese, every white duck might be an Aylesbury and every white goose an Embden.

White domestic geese have long been popular in many parts of Europe. They are ornamental, like swans, and they also have practical qualities. White birds with pale skins provide an attractive carcass. Undoubtedly, though, the live birds can be spotted more easily by the gooseherds, and distinguished from the camouflage of the wild greylags. Whatever the reason, white geese have been bred from the North Sea to the Black Sea: large ones from the coastal regions of Germany (the areas near Emden and Bremen); small ones from central Europe (Italy and Bohemia); curly ones from the Danube basin and the Black Sea coasts. White geese are even mentioned by historians in connection with the siege of Rome by the Gauls in the fourth century BC.

It would be fair to say that the genetic mutations that led to the breeding of white greylags became part and parcel of the domestic goose 'industry' across Europe. The so-called white breeds are probably little more than local 'strains', traded and crossed across the continent. The more distinctive forms have acquired 'breed' status and many have been described and classified in 'Standards of Excellence'. Whether all the Roman geese actually came from Rome, or just the Romantic imagination of some goose dealers, remains in doubt. Certainly, if the tiny white Czech geese (recently standardized) become popular, it is likely that the two breeds will become inextricably crossed.

Common characteristics of all white geese include blue eyes and orange bills. Legs and feet tend to be orange as well, although an element of pink is frequent and may be related to the particular 'strain' of greylag. The eastern varieties have more distinctly pink bills and legs than the orange of their western cousins.

The white plumage is not quite as simple as it would seem. Ducks with white feathers tend to have yellow ducklings: they have a 'colourless' gene which takes away the pigment. Asiatic white geese are similar to the ducks, whilst European geese are white for different reasons. They have both dilution and 'spot' mutations that together allow a certain amount of colour in the baby fluff. Females are often darker than males, and young adult females can have some grey plumage, especially on the rump. The grey feathers in the females are often replaced by white ones in the next moult, so don't worry about them!

In Britain today it is possible to buy white commercial geese (traded under various names) as well as more stabilized 'pure' breeds: White Toulouse and Pomeranian are rare colour variants of more common wild-colour breeds. They have the same distinctive shapes as their coloured varieties. The four breeds described over the next few pages are the breeds that are most commonly shown in waterfowl exhibitions.

fact file

Classification	Heavy
Weight range (male)	12.7–15.4kg (28–34lb)
Weight range (female)	10.9–12.7kg (24–28lb)
Colour	White
Wild ancestor	Greylag
Origin	North Germany
Status	Rare (UK)
Eggs	Up to 30

EMBDEN GEESE

The first British 'Standards' of 1865 included just two breeds of geese: the Toulouse and the Embden. They were also in the first USA Standards, 1874. The Embden – initially called Bremen by Americans in 1820 – was imported from the northern coastal areas of Germany. Records of goose breeding in this part of Germany date back to the 13th century.

These birds were big, significantly bigger than the 'Common' geese in Britain and the USA at the time. Harrison Weir (1902) refers to birds of 'much over thirty pounds' (about 15 kg) which he knew as 'Brunswicks' 40 years earlier. Edward Brown wrote in 1929 of birds of this type bred on the plains of Hanover, Westphalia and Oldenburg. He added that 'the present-day Embden is the result of crossing the German White and the English

White geese'. Certainly, exhibition Embdens in Britain are more solid and chunky than the swan like geese now shown in Germany. American birds also tend not to be quite as heavy as British ones and stand taller.

Pure Embdens tend to lay early, often by mid-February, and generally produce about 30 eggs, each weighing 170g (6oz) or more. The birds are rather too large to incubate their own eggs, unless they have a very large and deep nest; there is a real risk of them squashing their eggs and killing their hatchlings. Some strains are very tame if hand-reared, though their size makes them imposing. Commercial white birds mistakenly thought to be Embdens, are smaller but can be a problem in the breeding season.

The plumage of the Embden is much tighter

than that of the Toulouse. The bird also has a more erect carriage and a smooth breast, and there should be no sign of keel or dewlap. Both bill and shanks are bright orange. Most Embden goslings can be sexed according to down colour: females are darker grey (in a saddleback pattern) than the males, a difference that is evident until two or three weeks of age.

Much of the commercial goose industry in Britain is based on crosses between the Toulouse and the Embden. These crosses can themselves eventually breed pure white geese, which are often found in auctions as 'Embdens'. If pure Embdens for exhibition are required, then find an established breeder with a track record.

ROMAN GEESE

fact file	
Classification	Light
Weight range (male)	5.4–6.3kg (12–14lb)
Weight range (female)	4.5–5.4kg (10–12lb)
Colour	White
Wild ancestor	Greylag
Origin	South and Central Europe
Status	Critical (US)
Eggs	40–60

Italy is not renowned for its goose breeding in modern times. Geese are well endowed with insulation in the way of feathers and down and tend to migrate north in the warmer months, so central and southern Italy are not likely areas for goose breeding. The colder regions, such as Lombardy, Piedmont and Venezia, are perhaps better suited. Edward Brown identifies two strains of the same 'race': the Padovarna, which has elements of grey in its feathers, and the white Roman. He also says, 'Throughout South Germany, Austria, Hungary, and the Balkan States, is found a white goose, partaking of the Embden in character, but much smaller in size, and resembling the White Roman goose.'

The difficulty with most white breeds is being certain of their origins. Also, buyers are less willing to pay reasonable prices for them. Toulouse and African geese fetch twice as much (at least) as top-quality white geese such as Romans or Embdens.

Nonetheless, Roman geese are beautiful little creatures. They, and the even smaller Czech geese, have the same appeal as Call ducks. They are small, chubby, easy to keep and good breeds to start with if you have limited space. Some strains can be very placid; others quite aggressive, but this probably says more about their upbringing than their genetics. Breeding hormones make a difference, but by far the biggest variable is the human one: tame geese tend to come from careful breeders.

Both Czech and Roman geese can manage limited sustained flight thanks to their small body weight and large wing span. When moved to new premises, or on a windy hill, they need to be clipped on one wing (see page 37).

Romans and Czechs are good layers and lightweight sitters. They often lay over 40 eggs a year; as many as 60 are possible. The eggs, of course, are much smaller than those of other white breeds, weighing around 140g (5oz).

CZECH GEESE

fact file	
Classification	Light
Weight range (male)	5–5.5kg (11–12lb)
Weight range (female)	4–4.5kg (9–10lb)
Colour	White
Wild ancestor	Greylag
Origin	Bohemia
Status	Recently standardized (UK)
Eggs	40–60

Small white geese from south-east Germany and the neighbouring Czech Republic, specifically the old kingdom of Bohemia, have become very popular in recent years. It is hard sometimes to tell the difference between

SEBASTOPOL GEESE

fact file	
Classification	Light
Weight range (male)	5.4–7.3kg (12–16lb)
Weight range (female)	4.5–6.3kg (10–14lb)
Colour	White
Wild ancestor	Greylag
Origin	Danube/Black Sea
Status	Rare (UK), threatened (US)
Eggs	30–40

Sebastopol, or Danubian, geese arrived in Britain shortly after the end of the Crimean War (1854–6) and were exhibited at the Crystal Palace in 1860. A newspaper engraving of the time shows a pair of 'smooth-breasted' geese, each weighing 11lb (about 5kg). They are not much larger than Roman geese, to which they may be related. What makes them special are the long, curled feathers which trail in ribbons to the ground.

The explanation for this unusual look is that certain feathers on the Sebastopol are elongated. The central stem is prone to splitting and the edges of the vanes tend to show waves or flutings. What causes this to happen is something we call a 'Sebastopol gene'.

Two such inherited mutations cause the plumage to be extremely 'curly'. The body feathers, right up to the lower neck, are curled; the wing feathers are so degraded that the bird is not capable of flight; the shoulder and thigh feathers are long and often wispy. The whole bird looks like a ball of fluff.

One inherited mutation, on the other hand, has a reduced effect, turning only the shoulder and thigh feathers to ribbons. The breast is smooth and the wings capable of flight.

If you have two such 'smooth-breasted' parent birds, they are likely to produce three sorts of offspring:

* 'smooth-breasted', like the parents
* 'curled-feather', like balls of fluff
* normal geese, like ordinary white Roman-types.

Only the last two will 'breed true'. The curly form was standardized in the USA in 1938 and there are beautiful examples of these birds in Australia too; in Germany only the smooth-breasted is accepted.

Some people find Sebastopol Geese unattractive. Others think they are fascinating and call them Pantomime geese. Contrary to what one might believe, they manage to keep as clean as other white geese, even in the worst weather, as long as they have paddling or swimming water. In the breeding season they tend to lose some of the soft feathers.

Sebastopols are productive, too. They lay 30–40 eggs per year in good strains and are good sitters.

Roman and Czech geese. Extreme examples of the Czech are very small. They have short bills and short necks, very chubby faces and neck feathers that stick out in a way reminiscent of German Pekin ducks. Their bodies are described as 'oval' or 'egg-shaped'. They have a reputation for stroppiness in the breeding season when they have goslings, especially with other waterfowl, although many breeders claim them to be no worse than most geese.

Czech geese have been standardized in Germany, and more recently (2008) by the British Waterfowl Association.

PIED GEESE

'Pyed' geese were first described in Britain by Gervase Markham in 1613. Pied is quite a common colour form and is found in many regional breeds throughout Europe. The birds are basically white with areas of coloured plumage on the head and neck, over the scapular feathers (that cover most of the wings, when not in flight, and give a heart-shaped mantle) and over the thighs.

The pied or 'spot' gene is found in white geese but not in whole-coloured birds. It is possible therefore to 'manufacture' grey backs and buff backs by crossbreeding over a number of generations, and this is what often happens with commercial Embden–Toulouse hybrids. The result is frequently a large, fast-growing meat bird with coloured patches. Perfect exhibition specimens with standard symmetrical markings may be quite rare.

The areas of Germany, Poland and Sweden that surround the Baltic Sea are well known for their pied geese, recorded in a painting by Jakob Samuel Beck (1715–78). Most famous is perhaps the Pomeranian; then there are minority breeds such as the Skåne, Öland and Flemish geese. How often they have been crossed with large white geese, such as the Embden, is debatable.

POMERANIAN GEESE

fact file	
Classification	Medium
Weight range (male)	8.2–10.9kg (18–24lb)
Weight range (female)	7.3–9.1kg (16–20lb)
Colour	Grey-and-white (pied), grey or white wild-colour white
Wild ancestor	Greylag
Origin	North Germany and Poland
Status	Critical (US)
Eggs	30–40

'Pomerania' stretches along the Baltic coast from Stralsund in the west to Gdansk in the east and includes parts of modern Poland and Germany. Its name reflects its position: 'land next to the sea'. Like much of northern Germany it has a strong tradition of goose-breeding and the culinary specialities of smoked goose breast, salted goose and goose fat may have affected the type of birds bred: geese with long bodies and prominent chests have traditionally been highly regarded.

Nowadays the key recognition point of the German *Pommerngans* is the single distinctive central fold in the abdomen. This is what makes its white variety different from the Embden and the pied version different from the English Grey Back. Quite large and with a stout pinkish-orange bill, the Pomeranian is a good utility and show bird. The pied form is most popular in Britain whilst the all-grey is commonly exhibited in Europe. As with all geese, behaviour depends on upbringing, but some strains of Pomeranians can be quite assertive. In a good strain, young females produce more eggs than Embdens.

In the USA, the breed was standardized in 1977. Although the Standard follows the German requirements, any saddleback geese in grey or buff tend to be called 'Pomeranian'.

GREY BACK AND BUFF BACK GEESE

fact file

Classification	Medium
Weight range (male)	8.2–10kg (18–22lb)
Weight range (female)	7.3–9.1kg (16–20lb)
Colour	Pied
Wild ancestor	Greylag
Origin	Northern Europe
Status	Well marked specimens rare
Eggs	Depends on the strain (see main text)

Grey Back and Buff Backs are identical in shape. They should be fairly large and their key requirements for show purposes include an even, double paunch and symmetrical, well-defined areas of wild-colour (grey) or buff, which should cover the head and roughly one third of the neck. Both breeds have patterns on the coloured feathers very similar to those of the greylag and the Toulouse: grey or buff with light edging.

The buff dilution is sex-linked. A Buff Back male bred to a Grey Back female will produce only Buff Back female goslings. Its male goslings will look like Grey Backs, but a proportion of their offspring will show buff plumage.

Creating healthy new Buff Back stock from scratch might involve mating an American Buff gander with a Grey Back goose. The 'spot' gene, forming the saddleback pattern, is recessive, so it will take several generations to produce well-marked offspring.

The number of eggs produced depends on the strain used for cross-breeding. They can be crossed with Embdens for size, or smaller geese for greater egg production.

GREY GEESE

It is not surprising that the one species of wild goose to be domesticated in Europe was the greylag (*Anser a. anser*). Out of all the species of grey goose, this is the largest and the least wary. Migrating north later than most other geese, greylags breed south of the Arctic Circle from Iceland to eastern Russia. They have large heads and substantial bills. Those of Western Europe have bright orange bills and tend to be slightly smaller than those from further east (*Anser a. rubrirostris*), which have paler pink bills and eye rings.

Over a period of 3,000 years or more, greylags have been bred in captivity, rather than just hunted in the wild. The most extreme result is the exhibition Toulouse, originating in southern France where there are several commercial grey varieties. In Alsace there is a small grey domestic goose, much closer in shape to the original greylag; there is also an all-grey version of the Pomeranian. The original Steinbachers had grey plumage, but they may have some Asiatic ancestry. The grey of these breeds is referred to as 'wild-colour' because it is the original colour, lacking other genes such as 'spot', 'dilution' and buff.

TOULOUSE GEESE

fact file	
Classification	Heavy
Weight range (male)	11.8–13.6kg (26–30lb)
Weight range (female)	9.1–10.9kg (20–24lb)
Colour	Grey (wild-colour), Buff or White
Wild ancestor	Greylag
Origin	France
Status	Watch (US)
Eggs	30–40

Toulouse geese became famous in the early 19th century, first as a speciality table bird and then as breeding stock. The Earl of Derby imported them into the UK in the 1840s and they were exhibited at the first National Poultry show in 1845. Twenty years later, Toulouse and Embden were the only goose breeds described in the first Poultry Standards. For much of the 20th century the Toulouse has proved useful as a premier show bird and as one of the parent breeds for the commercial goose industry.

The Toulouse has a very distinctive shape. In profile it looks almost rectangular. The back is horizontal, the body has a deep double paunch and suspended from the breast bone is the 'keel', a deep fold of skin and tissue.

Other key features include a short, very strong bill below which is a large, pendulous dewlap. The plumage is bulky and quite soft, often making Toulouse appear larger and heavier than they really are. Such soft feathers can cause problems in bad weather: the birds become cold, wet and bedraggled, much more than geese with harder, sleeker plumage. It pays to protect Toulouse from very wet weather, especially the Buff variety.

For such large birds, Toulouse geese are very shy and reserved. They are a bit cumbersome, and do not like being rushed or driven on steep ground. The active breeding life of exhibition stock is quite short. Large

Toulouse may be less active at six to eight years, and some strains develop twisted flight feathers. Commerical strains are less vulnerable. Large birds especially benefit from paddling water to keep themselves clean.

Exhibition Toulouse are strictly for serious breeders. There are also smaller utility types which are quite good layers (30–40 eggs). That is why they are crossed with Embdens to produce commercial geese.

BLUE GEESE

Although grey was the first Steinbacher variety to be standardized (in 1932), the blue is now the most popular. It is more than likely that crosses with Asiatic geese from Russia allowed the gene into flocks of these German birds.

The blue dilution gene in geese seems to work in a very similar way to the equivalent duck gene. It is not completely dominant and, fortunately, it is not sex-linked. Cross a grey Steinbacher with a pale blue one and the offspring will be blue, but quite dark. Further breeding of blue-to-blue will provide a proportion of light blue birds which together breed true. Both the grey and the light blue are stable colour forms. The Steinbacher is a much sought after breed in Britain and has now been introduced to the USA.

More recently, a further blue breed has emerged in Franken (Frankonia), Bavaria. The Germans call it the *Frankische Landgans*. In English this would be the Frankonian goose. There are also blue Chinese geese in Europe and the Far East.

STEINBACHER

fact file	
Classification	Light
Weight range (male)	6–7kg (13–15lb)
Weight range (female)	5–6kg (11–13lb)
Colour	Grey, blue
Wild ancestor	Greylag and swan goose
Origin	Germany
Status	Recently imported into UK and US
Eggs	5–40

In Germany the Steinbacher is referred to as a *Kampfgans* – a fighting goose. This is a relic of the goose-fighting that existed in Eastern Europe and Russia, much like cock-fighting in Western Europe. These birds are certainly some of the most confident with humans and make superb pets. They become very tame and are among the least aggressive. A small mixed flock of goslings will always seem to be mostly Steinbachers because they are the ones that push their way to the front and enjoy human contact, chewing clothes, hair and shoe-laces. It is the Toulouse and Sebastopols that seem to get left at the back.

Be warned: when Steinbachers are confronted with African ganders during the breeding season, neither breed will give in. They will not back off, or run away; they will fight until one is exhausted.

In addition to the blue feathers, Steinbachers have distinctive bills, quite short and very strong-looking. The upper line tends to be convex, not as extreme as that of the Russian Tula goose but reminiscent of it. The base colour is bright orange, but instead of having a pale bean on the end the Steinbacher has a black one. The serrations along the inner edges of the mandibles are also black, marked like lipstick. Young birds can all have signs of a dark line along the culmen of the bill. This often disappears in adult geese. Complete lack of black on the bill is likely to be a sign of out-crossing to other European breeds.

Steinbachers have never been bred for high production. Some strains lay very few eggs and have a short breeding life. Others lay well.

BUFF GEESE

Buff geese have the same plumage pattern as the Toulouse and the wild greylag. The only difference is the change of the basic colour from grey to buff, a yellowish beige. This is caused by a recessive dilution gene that is sex-linked: it shows immediately in females but can lie hidden in males. Cross a buff gander to a grey goose and all the female offspring will be buff; all the males will be grey. Yet in the next generation these grey males will father some buff offspring (the proportion depending on which colour mate they have and also on the sex of the goslings). It sounds complicated and makes you wonder how the breeds emerged by accident before anyone understood the genetics. However, because this is a recessive gene, it could have lain hidden in wild or domestic flocks long before it showed up enough to be noticed.

BRECON BUFF GEESE

fact file	
Classification	Medium
Weight range (male)	7.3–9.1kg (16–20lb)
Weight range (female)	6.3–8.2kg (14–18lb)
Colour	Buff
Wild ancestor	Greylag
Origin	Wales
Status	Rare (UK)
Eggs	Up to 30

In 1929 Rhys Llewellyn noticed what he called buff 'sports' in amongst a flock of grey and white geese in the Brecon Beacons in central Wales. Putting one of these females to a so-called 'Embden' gander (with no buff gene) he produced all-grey offspring. Sensibly Rhys Llewellyn then used a gander from the 1930 hatch (which had inherited one buff gene but did not show it) with two buff females from elsewhere. Several buff goslings were produced and these formed the stock which was reported to breed true by 1934.

Rhys Llewellyn was probably using the term 'Embden' loosely just to mean a white gander. The local flock of white and grey geese could easily have been 'common' or Pilgrim geese with auto-sexing characteristics. This would certainly explain the apparent lack of pied birds to emerge from his breeding programme.

The Brecon Buff has proved to be an extremely good goose for the smallholder: it is in great demand and not always easy to find in its pure form. Real Brecon Buffs have pale pink eye rims, bills, legs and feet, probably derived from the Eastern sub-species of the wild greylag.

True Brecon Buffs are quite chunky birds, slightly smaller and less 'rangy' than the American Buff. They also seem to benefit from good grazing, typical of the damp climate of central Wales. Unlike Toulouse or giant Embdens, they cope well with hilly ground and they are more economical to feed. Additionally they have a higher ratio of muscle to bone.

They lay between up to 30 eggs a year and also make good sitters and mothers. At 6–8kg (14–18lb) the females are less likely to break the eggs or squash the goslings than some larger breeds. It is still better to finish the eggs in the incubator and hand-rear the little ones until they are big enough to enjoy the company of a 'foster' parent. Some tame Brecons are excellent at this but quite often they will only accept buff goslings back. Our big African gander is perhaps the best of all: he is not 'colour prejudiced' and is extremely gentle. Some Brecon females drive off introduced goslings that are not the right colour. Hand-rearing may be safer, for the goslings.

On the whole, Brecons are quiet, well-behaved and easy to look after.

AMERICAN BUFF

fact file	
Classification	Heavy
Weight range (male)	10–12./kg (22–28lb)
Weight range (female)	9.1–11.8kg (20–26lb)
Colour	Buff wild-colour
Wild ancestor	Greylag
Origin	North America
Status	Rare (UK), critical (US)
Eggs	25–30

The American Buff goose was first standardized in the USA in 1947. It shares the same plumage pattern with the Brecon, although the feather colour tends to be slightly more brassy and less pink in the shade of buff. For shape and size the American more closely resembles the Embden but is taller, heavier and less compact. Both breeds are dual-lobed in the paunch.

There is some evidence that the American Buff came from Buff Back or 'Buff' Pomeranian geese imported from Europe. The greylag is not a native of the continent, so all the American domestic geese must have been imported at some time or other. These are generally amiable geese, like the Brecon Buff, with which it should not be confused.

Like the Brecon, it is a moderate layer. Production varies with age and strain.

AUTO-SEXING GEESE

Unlike many duck breeds, most geese are not 'auto-sexing'. The wild geese of the northern hemisphere lack what is called 'sexual dichromatism', meaning that you cannot easily recognize males from females by the differences in feather colour or markings. However, the ganders do tend to be slightly bigger; their body language is more self-confident and assertive, whilst the females have a deeper, gravelly voice. It is not easy, though, to be 100 per cent certain that a goose is female, unless you have seen her lay an egg; and just because one goose tries to 'tread' another, it is not necessarily a male. Geese without ganders can 'go through the motions'. 'Vent-sexing' (examining the sex organs) is the reliable method. In early feather, it can be easy to mistake one sex for the other.

However, there are exceptions to the 'sexual dichromatism' rule (if the birds are pure): e.g. the Pilgrim, the West of England and the Shetland, which is a smaller version of the West of England. There are also Normandy geese, Cotton Patch geese (USA) and Settler geese (Australia). The ancestors of these birds travelled the world at the time of European colonial expansion, when farmers took their stock with them to the New World.

Oscar Grow recognized the auto-sexing characteristic in birds he found in Vermont and other parts of New England. He called them 'Pilgrim Geese', he said, as a reference to his own pilgrimage from Iowa to Missouri during the Depression: research by Robert Hawes has produced no proof that geese came over with the Pilgrim Fathers. However, circumstantial evidence would suggest that the auto-sexing element has long been common in European farmyard geese, certainly in Britain and France.

Harrison Weir, writing in 1902, reports the preference amongst keepers of the 'Common' goose for white ganders 'even if the geese are grey. But this may be and is perhaps attributable to centuries of selection as to colour'. Jean Delacour recalls seeing the Pilgrim breed on small French farms, particularly in Picardy, at the end of the 19th century. Other reports note similar geese in Normandy. Oscar Grow may not have been the originator of the breed but he was certainly its most effective advocate. He was also an experienced geneticist.

It was F N Jerome's research in the 1950s that showed how the auto-sexing genes worked. The males have two 'dilution' genes, whilst the females have only one. This makes it what is called a sex-linked characteristic, like human colour blindness in reverse, where males need only one such gene to show colour blindness and females need two. The female goose can only have one dilution, so her wild-colour grey feathers are only slightly diluted to blue-grey. The pure Pilgrim male has a double dose, making him almost white.

PILGRIM GEESE

fact file	
Classification	Light
Weight range (male)	6.3–8.2kg (14–18lb)
Weight range (female)	5.4–7.3kg (12–16lb)
Colour	Grey female
	White male
Wild ancestor	Greylag
Origin	Britain, USA
Status	Rare (UK), critical (US)
Eggs	25–35

Pilgrim females have diluted grey feathers over most of the head, neck, body and wings. There is a hint of white on the face, usually around the eyes (forming spectacles). Some strains develop more white on the front and top of the head as they age, as well as specks on the neck. This is quite common but might be regarded as a slight defect. What should be avoided are white feathers on the wings. White flights are a sign of impurity.

Males are mainly white except for some light grey on the lower back and even a little on the secondary or tail flights. How else can you tell a pure white Pilgrim from a small Embden or a Roman? What you do not want is any white gander crossed with a pure Pilgrim female. Getting rid of the faults in future generations can be a nuisance: white flights on females and patches of white on the breast.

Pilgrims were first standardized in the USA in 1939, and the UK followed in 1982. They have since been recognized (from imported stock) in

Germany, and from Settler stock in Australia. Pure Pilgrims are quite rare. They are listed as a rare breed in the UK, together with the West of England and the Shetland. In the USA, their numbers (along with the Cotton Patch and Shetland) are critical.

Pilgrim geese benefit from good grazing and do not require a lot of high protein supplements (unlike some of the larger birds). Also, being quite light, they can sit a clutch without too much fear of squashing the eggs or treading on the goslings. They have the reputation of being a calm, sweet-natured breed and make good pets.

SHETLAND

fact file	
Weight range (male)	6–7kg (13–15lb)
Weight range (female)	4.5–6kg (10–13lb)
Colour	Grey-and-white female; white male
Wild ancestor	Greylag
Origin	Shetland
Status	Rare (UK), critical (US)
Eggs	Up to 40

A hard life on the islands beyond the north of Scotland may have favoured the Shetland's small size. It is a hardy bird with a neat, round body and well-developed wings. In 1989 S H U Bowie described how these geese were kept in the Shetland Isles, where a small population still remains. Like the Pilgrim and West of England, this is classified as a rare breed. As with all geese, behaviour depends on upbringing, but like the Romans and the Czechs, this is a useful breed for a small space.

WEST OF ENGLAND GEESE

This is a more substantial bird than either the Shetland or the Pilgrim. Not much is known about its genetics, nor those of the Shetland and Cotton Patch, which have a similar colour pattern, probably determined by the same genes. At first glance, the males look like Pilgrims (perhaps a bit whiter) and the females are superficially like Grey Backs. On the whole the grey plumage is not quite as dark as that of the Grey Back; the patches of colour are more irregular and asymmetrical, and sometimes there are no markings on the head and neck, especially as the females age. Despite its ancestry, the breed was only standardized in the UK in 1999.

fact file	
Classification	Medium
Weight range (male)	7.3–9.1kg (16–20lb)
Weight range (female)	6.3–8.2kg (14–18lb)
Colour	Grey-and-white female; white male
Wild ancestor	Greylag
Origin	England
Status	Rare (UK)
Eggs	20–30

Ducks

For thousands of years wild ducks have been hunted for their meat, their feathers and their eggs. Domestication may have begun earlier in the Far East, but it is only in the last few centuries that Europeans have made use of ducks bred for specific purposes. From chance mutations to deliberate selections, these are the birds which take centre stage at waterfowl shows, farmyards, duck ponds and back gardens – big ones, small ones, fat ones, thin ones – dozens of breeds and lots of colour varieties. The only problem is making the choice!

Domesticated Ducks

At some stage in prehistory, probably in many separate parts of the world, it was found to be useful to keep adult ducks in captivity. For later settled communities, this began the process of domestication. Just one species of wild duck provides the ancestry for nearly all domestic ducks – the wild mallard (*Anas platyrhynchos*). Around most of the cooler regions of the northern hemisphere this species is common, quite large compared to other dabbling ducks and less shy than most.

In Central America a similar process must have begun with the wild Muscovy (*Cairina moschata*).

Basically, domestication means selecting natural mutations and deliberately breeding those which benefit the human owners. Typical mutations would involve changes in:

- size
- shape
- behaviour
- growth rate
- breeding season
- plumage colour

Domestic animals tend to be bigger than their wild ancestors, often with more muscle on desired parts of the body. They are more placid, less likely to move great distances (either by flying off or by migrating) and more human-friendly. If they breed all the year round or lay large numbers of eggs, that is all to the good. If they are easily noticed in the field they are more easily herded or protected. Isolated by distance and historic limits in transportation, different parts of the world evolved quite different domestic duck breeds.

Rouens and Aylesburys

Until the late 18th century most European ducks were probably little more than 'puddle ducks', a miscellany of sizes, shapes and colours, more or less closely related to (and interbred with) wild mallards. Bigger commercial ducks were developed successfully at the time of the Industrial Revolution, when large numbers of country people were drawn to the towns and cities for employment and housing. Mass duck production (to supply the growing population of London) led to the growth of the Aylesbury

LEFT *American Pekin ducks.*

duck phenomenon (see page 71). The Aylesbury was a large duck, of a very similar size and shape to the French Rouen, but with white feathers and pink skin. It also bred at a slightly different time of the year and provided a lighter, more attractive dressed appearance. The white feathers were in higher demand than the wild-colour Rouen.

Pekins

In China there developed a completely different style of meat bird which we call the Pekin duck. It was eventually imported into the West around 1872–3, but only after hundreds of years of separate development under domestication in the Far East. This bird too was white, fast-growing, large and unlikely to fly. It could be bred and raised in great numbers and was ideal for supplying the needs of growing towns and cities. It was also unusual in shape, being very upright and fluffy.

Indian Runners

Further south, in Indo-China, Malaysia and Indonesia, similar upright ducks were developed but with a different priority – eggs. The famous Indian Runner (not from India, but from the islands around Java, Lombok and Bali) may have evolved over hundreds of years in that part of the world. By the time the Dutch and other sea traders 'discovered' it in the middle of the 19th century, or perhaps earlier, it was a prolific egg-layer. Instead of the dozen or two eggs typical of the wild mallard, some of these ducks could lay well over 200 in a year. The laying season was not limited to the spring, and the Runners could not fly off. They walked or ran everywhere, foraging in the rice fields, even walking all the way to the coast for market or export. They also had a range of interesting and attractive plumage colours.

CHOOSING A BREED

Like domesticated geese, most domesticated ducks are descended from the same common ancestor (in this case the mallard), but there is more variety of temperament between breeds of ducks than between breeds of geese: individual breeds – and strains within breeds – differ widely in the number of eggs they lay and the amount of noise they make. There are more details about this on pages 44–45 and in the individual breed entries in this chapter.

Nonetheless, the same basic rule applies with ducks as with geese: good early handling is the single most important thing in establishing a bird's temperament. So always buy from a reputable breeder – see page 32 for advice on how to find one – and get an idea of how the birds behave before you choose them. And do please read the chapter on Getting Started carefully before you make any decisions.

NEW BREEDS FROM OLD

The Rouen, Pekin and Indian Runner form the ingredients of the revolution that took place almost immediately the Asiatic breeds were brought to Europe. Rouen (or Aylesbury) crossed with Pekin provided the genetic material for the new table breeds. Rouen crossed with Indian Runner provided the egg-layers or general-purpose breeds. It is almost as simple as that.

Indian Runner
1.4–2.3kg (3–5lb)

Rouen
4.1–5.4kg (9–12lb)

Pekin
3.6–4.1kg (8–9lb)

Egg-layers
Campbell
Abacot Ranger
Welsh Harlequin
2.2–2.5kg (5–5½lb)

Table Breeds
Rouen Clair
Saxony
Silver Appleyard
3.2–4.1kg (7–9lb)

TRADITIONAL BREEDS

ROUEN

fact file	
Classification	Heavy
Weight range (male)	4.5–5.4kg (10–12lb)
Weight range (female)	4.1–5kg (9–11lb)
Utility	Table
Colour	Mallard
Origin	France
Flight potential	No
Status	Rare (UK), watch (US)
Eggs	35–100

Historically linked with Normandy – the area around the ancient city of Rouen – this breed of duck has plumage colour and markings very similar to those of the ancestral wild mallard. Because of the proximity of the southern coast of England, it is likely that breeding stock was traded between the two areas. As a result of British enthusiasm for this French duck, even the French started calling it the English Rouen (or *Rouen foncé*, dark Rouen). Imported to the USA around 1850, and exhibited as the 'Rhone duck', it was standardized as the Rouen in 1874.

The modern exhibition bird is probably not as fast-growing as those early Rouens. It is very large, heavy and ponderous, and should be calm and tame if it has been reared well. The back is horizontal, like that of the Toulouse goose, and from the chest a deep keel hangs almost to the ground. In silhouette, it closely resembles the English Aylesbury. The genuine exhibition Rouen duck has an orange bill with a dark saddle. The feather ground colour of the female should also have a bright chestnut or golden brown appearance, not a muddy or drab colour. Crossbreeding with Aylesburys can spoil the show appearance of these birds.

However, it is the Rouen that has provided the breeding stock for many new and vigorous domestic breeds. Crossed with the Pekin it provides the light phase plumage of the Rouen Clair. With added blue dilution from the Pommern, it produces the Saxony; and with another mutation from the Pekin it can produce the Silver Appleyard. It is the grandfather of the Campbell series and finds itself with its own Blue and Apricot colour forms.

This is not the best breed to start with. Rouens consume large quantities of expensive food, and such is the need for a high-protein diet in growing ducklings that they can have a tendency towards 'feather picking'. Life expectancy is lower than many lighter breeds. For exhibition breeders, Rouens are a great challenge. They are beautiful birds and do well in shows; yet, like the Toulouse geese, they become old and ungainly in a few years. They benefit from flat ground and a shallow pool, and need exercise to keep them fit. Rouens bred for production are fitter and may lay more eggs than true breed birds, which can produce around 100.

AYLESBURY

fact file	
Classification	Heavy
Weight range (male)	4.5–5.4kg (10–12lb)
Weight range (female)	4.1–5kg (9–11lb)
Utility	Table
Colour	White
Origin	England
Flight potential	No
Status	Rare (UK), critical (US)
Eggs	35–100

Descriptions of the duck industry in the Buckinghamshire town of Aylesbury began shortly before 1800, at the height of the Industrial Revolution. Even before that the town had a duck-keeping tradition. 'The poor people of the town are supported by breeding young ducks; four carts go with them every Saturday to London,' wrote Dr Pococke in the 1750s. As the new century progressed, larger numbers of ducks were sent by train. They were grown on farms, smallholdings and in cottage gardens, not in modern industrial breeding factories; but it was big business by the standards of the day.

By the middle of the 19th century, exhibition Aylesbury and Rouen ducks had the same prestige as Embden and Toulouse geese. They were the breeds of status at the first National Poultry Show of 1845, and they featured in the Standards of 1865. Yet by the end of the century the traditional duck industry had begun to decline. Whether it was due to the complacency of the duckers, their inability to modernize or simply that the ducks had become inbred or were being replaced by the new super-breeds (Indian Runner or Pekin crosses), the Aylesbury was slowly losing its pre-eminence. The new large-scale duck-rearing industry was developing in the eastern counties.

Today, few ducks are grown in the Aylesbury area, although one family has bred them for generations. As many as 10,000 of these large, pink-billed, traditional commercial Aylesburys are still reared each year for the specialist table market, and the rate of growth is such that they are ready for 'processing' in about eight weeks.

Not every large white duck is an Aylesbury. The true Aylesbury is a bit special. The long bill is a beautiful shade of pink – the colour of a lady's finger-nail' (J K Fowler in Lewis Wright) and the plumage is like satin in texture and appearance. Like the Rouen in many ways, the Aylesbury is now largely the preserve of the specialist exhibitor. There are still a few blood-lines left, but not many. Population reduction and inbreeding are the present dangers, and breeders go to great lengths to obtain stock from abroad to maintain the breed. Even in 1890, Lewis Wright mentioned the difficulty of obtaining Aylesburys of pure stock 'since the wide dissemination of the Pekin duck'.

Heavy breeds like the Aylesbury do not fly, so wing-clipping is not necessary. However, because of their low keel, long body and sheer size and weight they need a shallow source of water in which they can bathe regularly. This also aids mating, as their size means that they find it difficult to mate on dry land.

Aylesburys will eat as much food as you give them, so it is important to make sure that adult birds do not become overweight. Ideally they need to be kept in an environment where they can be active, otherwise they will have a shorter life expectancy. For the best fertility Aylesburys need to be fit, preferably free-ranging on pasture. Egg production is similar to the Rouen (see page 70), and like the Rouen they should be calm and tame if they have been reared well.

INDIAN RUNNER

fact file

Classification	Runner
Weight range (male)	1.6–2.3kg (3½–4lb)
Weight range (female)	1.4–2kg (3–4½lb)
Utility	Eggs
Colour	Numerous stable colour varieties
Origin	Indonesia
Flight potential	Poor
Status	Watch (US), popular (UK)
Eggs	Over 200 in good strains

Shortly after they were imported into Britain in 1835 by the 13th Earl of Derby, these ducks were given the name 'Penguin Ducks', because of their upright stance. They laid lots of eggs, they didn't fly and, in their native islands of Indonesia, they were allowed to wander around the fields picking up invertebrates and waste grain. Salted, the carcasses and the eggs were exported on sailing ships. These birds played an important role in the duck economy of South-East Asia and still do.

The Runners became famous in Europe and America in the 1890s following the publication of John Donald's pamphlet *The India Runner*, from which the breed got its popular name. Early colour forms included fawn, pied, white, grey and blue, but only the pied form was standardized in the USA (1898) and the UK (1901). By 1901 many Runners had been crossbred with local wild or domestic ducks. It was not until 1909, when Joseph Walton began to import birds direct from Lombok and Java, that the breed's original shape and other characteristics were regained.

The importance of Indian Runner ducks cannot be over-stressed. Runners brought in high egg productivity and a range of unusual colour mutations, dusky, blue, brown, and pied genes among them. These have had a profound influence on the breeding of most 'light' ducks, from Khaki Campbells to Welsh Harlequins. Pure Runners have now been standardized in 14

different colours in the UK and similar colours can be found in the USA, Australia, Germany and South Africa, making the Indian Runner a popular breed for keen exhibitors and pet-keepers alike.

Indian Runners are generally very easy to look after. Although they do not lay as well as the Campbell, they can produce over 200 eggs a year in good strains. Excellent at slug control, these active birds spend more time foraging and less time on the water than other breeds. Just a few words of warning: firstly, don't expect the females to make nice little nests and then incubate their eggs. Some do, but most just dump the eggs in the middle of the field and wander off, leaving the eggs to the crows and magpies. The good thing is that most of them lay before 8 a.m., so if you don't let them out of their overnight shed until after 8.30, you can just pick up the eggs. If you want to hatch them, you will have to find broodies or to buy a decent incubator. Feed laying Runners special layer pellets (see page 40).

The other warning is about drakes. A pair of birds is usually a satisfactory arrangement. One male with a group of females is fine, as long as he does not confine his attention to a selected female. What is not acceptable is a gang of drakes. A female Runner is very susceptible to 'going off her legs' in the laying season, which becomes a signal to the 'gang' that she is available for their not-so-amorous intentions. They may crowd round and squash her, rupturing the egg passage, causing prolapse and even drowning her in their sexual frenzy. In the breeding pen you need to follow the rule: 'One drake good; more drakes bad.' Keep a pen just for drakes if necessary. The females will be eternally grateful and, at least, undamaged.

BALI

fact file	
Classification	Light
Weight (male)	2.3kg (5lb)
Weight (female)	1.8kg (4lb)
Utility	Eggs
Colour	Various
Origin	Indonesia (Bali)
Flight potential	No
Status	Recently re-made by cross-breeding Crested ducks with Indian Runners.
Eggs	100–200

Crested Runner ducks from the island of Bali were imported into Britain during the 1920s when a Miss Chisholm and Miss Davidson also obtained more Indian Runners from Lombok and Java. These specimens were a valuable addition to the pure blood stock brought in by Joseph Walton in 1909. The white Bali duck photographed for *The Feathered World* in 1925 was very close in shape to the contemporary Indian Runners, a little chunkier perhaps and with not quite such a straight bill. It was nonetheless a true Runner, with a small, globular crest at the back of its crown. The modern Bali's behaviour is similar to that of the Indian Runner (see opposite), and the Bali crest is the result of the lethal gene described on page 85.

It is doubtful whether any examples of the original line have survived, though recent crosses between Indian Runners and Crested Ducks have recreated the same breed form.

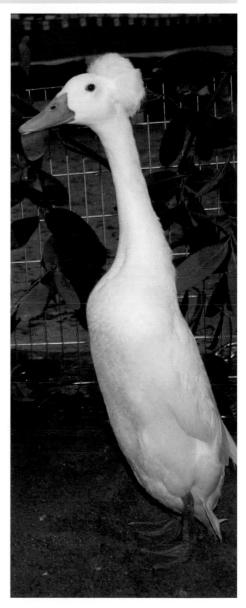

PEKIN

Pekin is the old name for the capital of China, Beijing. Ducks have been bred in the area for over 2,000 years, and they have become traditional items of Chinese cuisine. When first spotted by some westerners, these ducks were thought to be as large as small geese. They were imported into Britain and America in 1872–3.

Upright and 'cuddly', Pekins have chubby cheeks, short bills and 'smiley' faces. Their profuse feathering gives a false impression of body shape. The bushy neck feathers, in particular, create the illusion that they have little or no neck compared to other ducks. When their plumage is new and clean, and they have been feeding on the right food, they are wonderful-looking animals and are almost 'canary yellow' in colour, a characteristic demanded by early Standards.

The Pekins bred in Europe, especially in Germany, are very upright and close to the 'Teddy bear' image, with a body like a little boat standing on its stern. The American version looks more like a cross with an Aylesbury, which it may well be. The Pekin–Aylesbury cross made a massive impact on the table duck industry on both sides of the Atlantic. James Rankin, the 'father' of the Pekin duck industry in the USA, developed the commercial Pekin. Hybrid vigour helped to combat any inbreeding that may have reduced the Aylesbury's pre-eminence. It also allowed further experimentation with 20th-century 'designer' breeds.

White ducks are white because they have a genetic mask of recessive 'colourless' genes. Crossed to coloured breeds they immediately lose the whiteness and reveal the coloured plumage that lies hidden beneath. Black, bibbed, restricted mallard and light phase genes may all be found. C S Th Van Gink, artist and an authority on poultry in the Netherlands, intimated that the French Rouen Clair may have derived from a cross between the Rouen and the Pekin. He was probably right. A blue version of the Rouen Clair is called the Saxony duck, and that was certainly the product of interbreeding Rouen, blue and Pekin ducks. The Silver Appleyard probably arrived by a similar route.

All of these ducks are large, about the same size, and semi-upright, halfway in carriage between the Rouen and the Pekin. All three of these ducks emerged after the importation of Pekins into Europe.

The Pekin continues to do well in shows and exhibitions on both sides of the Atlantic, with the pure-bred German type being particularly popular in the UK although it is not as good a layer as the American. For management, the Pekin's fluffy feathers do have a down side. Sleeping birds lie stretched out like 'dead ducks'. They have to: their ruff of neck feathers makes it difficult to place the head over the shoulder. It certainly looks odd. They must be kept clear of parasites, notably the northern mite. Also, they suffer from mud, particularly around the eyes. Ready access to water is required, especially in hot temperatures when they are apt to overheat and dehydrate. Give them plenty of shade and keep them clear of parasites.

HOOK BILL

fact file

Classification	Light
Weight range (male)	2–2.25kg (4½–5lb)
Weight range (female)	1.6–2kg (3½–4½lb)
Utility	Eggs
Colour	Dusky mallard, biibed or white
Origin	Unknown
Flight potential	Possible
Status	Rare (UK)
Eggs	100–200

Compared to the wild mallard, most of the traditional domestic breeds look odd, or at least different. The Hook Bill is no exception. With its elongated bill with a semi-circular downward curve, it looks like a creation of fantasy. It is small and known to be a good flier – clip it if shows a tendency to fly off (see page 37). A group may stay together in a little clan. Hook Bills are quiet and unobtrusive, but also extremely inquisitive and quite tame. The young have been known to form a circle around visiting humans and tug at shoe-laces. As pets, they are underestimated.

Where they come from no-one seems to know for certain. Assertions that they originated in the Far East are probably guesses. Their popularity in Holland might indicate a connection with the Dutch East India Company which traded extensively in South-East Asia from 1602 to 1800. They were reported by Francis Willughby in 1678, and by William Ellis in 1750 as the Crook Bill duck: this was a time when the Aylesbury was not in full production and the 'Rouen' was only just becoming fashionable with the 'gentry'.

Willughby's had a mallard-like plumage. His duck had eye stripes, unlike modern Hook Bills. This may indicate the original colour type but it could also suggest crossbreeding with mallards. Modern Hook Bills have the same basic plumage as the Dark Campbell (dusky with no dilutions). A white bib is a common variation, and there are also pure white forms. Some people think that the white bib distinguished the Hook Bill from the mallard in flight so that they would not be shot by local wild-fowlers – a good story.

Once bred intensively in parts of north Holland and Germany, the Hook Bill is now quite a rarity and is mainly of 'ornamental' and pet value. The extreme shape of the bill seems to be of no disadvantage for foraging.

For those wanting a quiet, economical duck, that lays well and does not cost a fortune to keep, the Hook Bill is an excellent choice. It is perhaps the oldest pure breed of domestic duck in Europe and worthy of conservation. It is a rare breed, only recently introduced to the USA.

CAYUGA

fact file

Classification	Heavy
Weight (male)	3.6kg (8lb)
Weight (female)	3.2kg (7lb)
Utility	Table
Colour	Black. Early eggs black coated.
Origin	USA
Flight potential	No
Status	Threatened (US)
Eggs	100–150

The Cayuga's plumage is what makes it spectacular. The pigment that covers the feathers makes them black, but structural qualities of the feathers themselves intercept and scatter the light to give a brilliant green lustre. The sheen changes with the direction of the light and the position of the viewer.

Early reports of the duck near Lake Cayuga in North America indicate a large bird with black plumage. It has taken years of domestication to perfect the colour form and produce a bird fit for the table and the exhibition pen. Guesses are that the colour form emerged from the American Black Duck (*Anas rubripes*), a close relative of the northern mallard with which it will breed quite readily. The problem is that the Black Duck is not really black at all, just a dark dusky with little difference in plumage between the sexes; scientific analysis has not yet determined a strong link to the Cayuga or the little Black East Indian duck.

American records of black ducks go back to about 1809. There are reports of large black ducks with white bibs in England in the 1840s, and the French have a breed, the Duclair, also recorded in the 19th century, with blackish feathers and a white bib. Whether these breeds are related, there is no real evidence. We know that William Cook managed to cross Indian Runners, Rouen and Cayuga ducks in the 1890s to produce not only his famous Buff Orpington but also blue and black forms (with bibs). The Blue Orpington closely resembled the German Pommern and the so-called Blue Swedish, both with blue feathers and a white bib (plus white wing tips in the case of the Swedish).

Black plumage is not ideal for marketing table birds, and commercial Cayugas in the USA were replaced by the white Pekin after 1873. The Cayuga is now a popular exhibition bird valued for its striking plumage; the breed was in the first Standards of both the USA and UK. It lays eggs coated with a black cuticle. As more eggs are laid, they simply look white (or green).

Like all larger ducks, Cayugas are good foragers and fairly placid if reared well.

SPECIALIST BREED

MUSCOVY

fact file

Classification	Heavy
Weight range (male)	4.5–6.3kg (10–14lb)
Weight range (female)	2.3–3.2kg (5–7lb)
Utility	Pet
Colour	Various stable colour forms
Origin	Central and South America
Flight potential	Yes
Status	Widely available
Eggs	70

Since its discovery by the Spaniards in Central and South America, before 1500, the Muscovy has been exported all over the world. It is a different species from the mallard. Crossed with other domestic ducks, it forms a vigorous and fast-growing but infertile hybrid, a 'mule'. As a meat bird, it is more likely to be found labelled as a 'Barbary Duck' on the supermarket shelf.

The wild Muscovy (misnamed, like Albin's 'Moscovian Goose') is one of a group of perching ducks, such as the white-winged wood duck and Hartlaub's duck. The wild plumage is essentially black, although some of the wing coverts are often white. Red caruncles (fleshy growths) adorn the faces, especially of the males, and they have erectile fore-crown crests. These are quite heavy birds, with short legs, broad wings and a horizontal 'carriage'. They are excellent fliers, though males may be grounded by their weight; females and lighter drakes need to be clipped (see page 37). They also enjoy perching, which is unusual for domestic ducks. Look at the claws and webs of the Muscovy foot and you will see something equally well developed for perching as for swimming.

Domestication came before the birds were discovered by Spanish invaders. There were white Muscovys on Guadaloupe, reported by Diego Alvarez Chanca, in 1494. The domesticated varieties tend to be, understandably, bigger than the wild form. They also have more highly developed caruncles, which may not be beneficial for the birds as they get in the way of their eyes. Some countries now try to discourage the showing of drakes with too much in the way of facial caruncles.

Domestic colour forms are numerous, but the colour genetics do not follow the same rules as for domestic mallards. At the moment, standardized forms include wild-colour (Black), Blue, Chocolate and Lavender. These are all plain colours with possible white patches on the wings that may also develop with age. There are Magpie forms in the same colours (with caps, mantles and stern markings) and also pure White. American and European descriptions include White-headed forms and other rare mutations.

They are mostly reared as table birds. Australian birds, in particular, are bred for size.

Muscovy females, which are significantly smaller than the males, lay around 70 large, white eggs. They make excellent broodies and mothers; the incubation period is longer than for other ducks, 35 days. Both males and females make little noise.

They are good foragers and need protein in their diet. In the wild, they can catch crabs, fish and termites. Some drakes should not be kept with other domesticated ducks. Watch their behaviour: they can harass females.

MODERN TABLE BREEDS

ROUEN CLAIR

fact file

Classification	Heavy
Weight range (male)	3.4–4.1kg (7½–9lb)
Weight range (female)	2.9–3.4kg (6½–7½lb)
Utility	Table
Colour	Light Mallard
Origin	France
Flight potential	No
Status	Main stock in France
Eggs	Over 150

In both German and French standards they use the term 'trout' to indicate the light phase mallard plumage. This is a very expressive term for the golden fawn (*isabelle clair*) colour with dark markings like those on the flanks of a brown trout. It is not the same as the wild mallard or the traditional dark Rouen. This mutation is found in ducks of the Far East, especially birds of the Pekin and Runner type. Key elements are the male's white stern and slightly ragged claret bib, each feather fringed with white; and the female's white eyebrows and throat, plus the single chevron on the pale feathers.

The Rouen Clair is a modern utility duck largely created by René Garry between 1910 and 1920. Starting with a selection of farm ducks from Picardy, M. Garry set out to increase the size and maintain the *clair* colour. It took him ten years, using various lines of birds, some of which contained Rouen-mallard crosses. The eventual result, described in the 1923 French Standard, weighed up to 4.5 kg (10lb).

The Rouen Clair seems to be more popular in France than in the UK where good examples which breed true are presently difficult to obtain. It should be a quick-growing utility breed.

SAXONY

The Saxony is basically the same bird as the Rouen Clair in weight, shape, carriage and plumage colour, except for a pair of blue dilution genes. These two genes change the black pigment into what the Germans call 'pigeon blue'.

This beautiful bird has only been around since the 1930s; Albert Franz exhibited his ducks in the Chemnitz-Altendorf show of 1934. They were the result of careful breeding experiments using blue Pommern ducks (to obtain the blue dilution) and crosses between Rouen and Pekins (for size, shape and light phase plumage).

The blue dilution has the effect of making the female's plumage a beautiful shade of apricot buff. All the pencilling is diluted, to the extent that it merges with the background fawn of the feathers. Other special features include almost yellow bills with pale beans. The drake's neck ring is expected to go all the way round the neck without a break, in most cases, and the female is preferred to have no 'dribbles' of white colour from the throat on to the chest.

The Saxony is a healthy hybrid. It grows faster than the traditional Rouen and puts on weight at an early age, making it an ideal table bird. The adults are less heavy and cumbersome than Rouens. They are active in the farmyard, producing up to 150 eggs per year, as long as they are not inbred. The Saxony makes an excellent show and utility breed. It was standardized in the UK in 1982, and introduced to the USA in 1984.

fact file	
Classification	Heavy
Weight range (male)	3.6kg (8lb)
Weight range (female)	3.2kg (7lb)
Utility	Table
Colour	Apricot Light Mallard
Origin	Germany
Flight potential	No
Status	Critical (US)
Eggs	Up to 150

SILVER APPLEYARD

fact file	
Classification	Heavy
Weight range (male)	3.6–4.1kg (8–9lb)
Weight range (female)	3.2–3.6kg (7–8lb)
Utility	Table
Colour	Restricted Light Mallard
Origin	UK
Flight potential	No
Status	Rare (UK), critical (US)
Eggs	150

Reginald Appleyard was a dominant figure in the mid-20th century. He wrote extensively on duck and goose breeding, won medals for his exhibition waterfowl and had a thriving business in Ixworth, Suffolk.

The large breed that bears Appleyard's name is a bit of an enigma. He does not mention it in his books and there is little information about its development. It was not standardized in his lifetime and almost disappeared until the final decades of the 20th century. In the 1970s, Abacot Rangers (Streicher ducks) were being imported and sold in Britain as 'Silver Appleyards'.

Appleyard does describe the breed very briefly in his advertising pamphlet (c.1947): 'An effort to breed and make a beautiful breed of duck. Combination of beauty, size, lots of big white eggs, white skin, deep long white breasts. Birds have already won at Bethnal Green and the London Dairy Show and ducklings killed at 9 weeks, 6½ lbs. [about 3kg] cold and plucked.' It is likely that Appleyard discovered these colour genes in crossbreeds from the Pekin. Photographs of market ducks in the Far East today show birds that are almost identical to Appleyards.

The modern Appleyard is largely the recreation of Tom Bartlett of Folly Farm in Gloucestershire. Armed with a print of a 1947 painting by Ernest Wippell of Appleyard's birds and some restricted mallards from a market, he set about reproducing the famous breed. He succeeded, producing not only a full-size version but also a miniature form, roughly a third of the size of the large Appleyard (see page 87). The large form reached the Standards in 1982: two for the price of one, and a fitting tribute to two great waterfowl specialists.

The utility characteristics mentioned above can be lost by inexpert breeding, as can size and the typical colour. That is probably why the Appleyard was not standardized earlier. Nevertheless, people love this striking breed. It is now exhibited in both the UK and USA.

BLUE SWEDISH

fact file	
Classification	Heavy
Weight (male)	3.6kg (8lb)
Weight (female)	3.2kg (7lb)
Utility	Table
Colour	Blue
Origin	Europe
Flight potential	No
Status	Watch (US)
Eggs	Up to 150

Blue (with black) plumage crops up in several breeds of duck in Europe in the 19th century, roughly at the time of the first recorded importation of Asiatic breeds. Pied (Runner) plumage mutations show up earlier in 17th-century Dutch paintings, so Indian Runners may have reached Europe with increased trade with the 'East Indies' by the Portuguese or the Dutch. No written evidence seems to have survived. Runners owned by Lord Berwick in the 19th century had dull blue and blue-black plumage. William Cook 'created' Blue Orpington ducks in the 1890s using Runners, Aylesburys, Rouens and Cayugas. The Cayugas furnished the black component; it is likely that the blue genes came from the Runners. By this time there were several blue breeds: Termonde, Merchtem, Huttegem, Pommern, Blue Swedish, even blue Lancashire ducks in 1860.

Today in Britain only the Blue Swedish is recognized in the Standards. The Pommern (Pomeranian) is roughly the equivalent in Germany. Both of these breeds come from the same south Baltic coastal region, once claimed by Sweden. They are large ducks with blue plumage and white bibs. The Swedish has also a couple of white primary feathers on each wing. Smaller breeds, like the German Gimbsheim and the Belgian Blue Forest, have no white feathers.

Genetically these birds are black, plus a single blue dilution. If you breed two of them together something almost magical happens: the ducklings occur in three colour forms – black, blue and pale blue-white. Black ducks mated to black drakes produce black ducklings. Pale blue-white to pale blue-white produce pale blue-white ducklings. However, when you breed black to pale blue-white, *all* the ducklings turn out to be a rich dark blue. This presents one of the steep learning curves in understanding duck colour breeding. Once you have grasped the idea, lots of more advanced concepts slip into place.

These are quite fast-growing, plump birds, more erect in carriage than heavy breeds such as the Rouen. Egg-laying capacity is similar to the Saxony at around 150 a year. They are attractive birds for exhibition, standardized in the USA as early as 1904. Only in 1982, after American imports helped the breed, did they reach the UK Standards. However, to get the perfectly marked bird, large numbers have to be bred, which does discourage duck breeders from keeping them

Black, chocolate and dark blue breeds can suffer from a developmental condition. As they get older, the females tend to develop more white feathers until they are almost colourless; this means that they are not showable, although you can still breed from them.

MODERN EGG-LAYERS

CAMPBELL

fact file

Classification	Light
Weight range (male)	2.3–2.5kg (5–5 ½lb)
Weight range (female)	2–2.3kg (4½–5lb)
Utility	Eggs
Colour	Khaki, Dark, White
Origin	UK
Flight potential	Possible but not likely
Status	Watch (US)
Eggs	Up to 300

Not the first of the light breed crosses from Indian Runners and Rouens – that honour goes to the Orpingtons – the Campbell ducks are nonetheless undoubtedly the pre-eminent light breed. In egg-laying trials during the 1920s and '30s, the Khaki Campbells won most of the awards. These were not legal trials, but attempts by various colleges and organizations to test which ducks laid the most eggs.

It all began when a doctor's wife, Mrs Campbell from Uley in Gloucestershire, wanted roast ducklings for her husband and son. All she had was 'a poor specimen' of an Indian Runner female (but one that laid 195 eggs in 197 days). This she mated to a Rouen drake and the rest, as they say, is history. Well not quite: her first offspring looked more like the modern Abacot Ranger (which was later developed from the Khaki Campbell). By mating the offspring to wild mallard and back to the Indian Runner, Mrs Campbell ended up with a light brown duck. She was disappointed. What she really wanted was a buff duck like that of William Cook. Hers, she said, 'would come khaki', so she called it the Khaki Campbell out of patriotism: the Second Boer War was still going on in South Africa.

All varieties of Campbell duck are 'dusky' rather than wild-colour mallard. The females (and both juvenile and eclipse males) have no sign of eye stripes. Other than the White variety, the birds have dark or drab-coloured secondary feathers, with no bright blue speculum. Under the wings the covert feathers are light grey rather than creamy white. The males have neither bib nor neck ring.

The Dark Campbell, announced in 1943, has the basic plumage colour of the mallard, other than those points mentioned above. The shoulder and body feathers of the female are light brown, pencilled clearly in a darker shade. The male has an even colour all over his body, whilst his head is a dark shade of green-bronze.

Like the Welsh Harlequins and the Abacot Rangers, Campbells are lightweight egg-layers. All three have a semi-upright stance, roughly half-way between that of the Runner and the Rouen. The head and body shape is similarly intermediate between the two parent breeds. The bill is fairly straight; the head slightly rounded but not heavy; the chest and abdomen more rounded than that of the Runner, but not enough to curtail the breed's active habits: Campbells are busy little ducks. They are good foragers and the females have a reputation for laying lots of white eggs.

Not every Khaki female will lay 200–300 eggs in a year – only those that have been bred for commercial or utility purposes. High-production layers must be fed the right diet (see page 40). Feed only layer pellets (no wheat) if egg shells are poor, and always make mixed poultry grit available. Being on free range and able to forage busily keeps the birds fit and they will live longer, producing large numbers of eggs for up to four years before production does decline. Campbells occasionally go broody but, of course, they are not renowned sitters: their job is egg production. Females alone can sometimes be purchased from commercial suppliers.

ABACOT RANGER

fact file	
Classification	Light
Weight range (male)	2.3–2.5kg (5–5 ½lb)
Weight range (female)	2–2.3kg (4½–5lb)
Utility	Eggs
Colour	Silver
Origin	UK
Flight potential	Unlikely
Status	Rare (UK)
Eggs	Over 200

The Abacot Ranger was produced from a cross between 'sports' from so-called 'pure' Khaki Campbells, mated then to a White Indian Runner drake. 'Sports' are usually the result of breed impurities, either from the wrong drake 'getting over the fence' or hidden recessive genes in a flock. Whether the 'harlequin' (or 'silver') colour form came from the Campbells or the White Runner is debatable. The result was a breed that looks most similar to early illustrations of the 'Original' Campbell.

The history of the Abacot Ranger begins simply enough: Oscar Gray of Abacot Duck Ranch, Friday Wood, near Colchester, started the process in about 1917, during the First World War. Whether because the war had pushed up the price of duck food or for other reasons, such as competition from other breeds, the Abacot Ranger did not become very popular, in spite of doing well in laying trials, notably winning a Silver Cup in the National Test of 1922–3.

After this success, the Abacot Ranger seems to have dropped below the radar, in Britain at least, for some years. Within Europe, it had a different history. In the middle years of the 1920s Abacots were shipped across the North Sea to Germany, where, largely thanks to the conservation efforts of a Herr Lieker, the stabilized Streicher was standardized in 1934. The Abacots in Britain today owe much to reimported blood lines and to the German written standard.

Abacots seem rare in Australia, but they have invented their own duck, the Elizabeth, affectionately called the Lizzie. It is a similar colour to the Abacot, smaller (at the bottom end of the light duck range) with an extra wash of colour. Rather strangely, as it was first developed as a small meat bird to suit local demand, it makes a friendly pet.

Abacot Rangers have all the practical virtues of the Khaki Campbell, with arguably a more attractive colour form. The females can be tame and approachable without losing any of the independence of good foragers. This is an excellent breed for those who want light pet ducks which clear slugs from the garden and provide a useful source of eggs.

WELSH HARLEQUIN

fact file	
Classification	Light
Weight range (male)	2.3–2.5kg (5–5½lb)
Weight range (female)	2–2.3kg (4½–5lb)
Utility	Eggs
Colour	Brown Silver
Origin	UK
Flight potential	Unlikely
Status	Rare (UK), critical (US)
Eggs	Over 200

This breed also emerged from a supposedly 'pure' flock of Khaki Campbells in 1949. It retained the brown genes of the original Khakis but was otherwise very similar to the Abacot Ranger and other 'Silver' varieties.

The first 'mutations', spotted by Group Captain Leslie Bonnet in Herefordshire, were given the name 'Honey Campbell', which was later changed to 'Welsh Harlequin' when the family moved to North Wales. The flock was founded on just two specimens, but Bonnet quickly built up a thriving business and sent ducks around the world.

He also crossed a Harlequin and an Aylesbury to produce what he wittily called the Whalesbury Hybrid. This was a heavier breed and some of its variants resembled the Abacot Ranger, with blue speculum but light-coloured bill.

In 1968 problems with the fox forced Bonnet to amalgamate his remaining Welsh Harlequins and Whaylesbury Hybrids, at some risk of spoiling the original blood-line. Fortunately, one of his previous customers, Edward Grayson from Lancashire, still had a small flock of Harlequins. To get back the vigour of the first Honey Campbell sports, Grayson introduced new Khaki Campbell blood. His reasoning was sound: the Harlequin was genetically almost identical to the Khaki Campbell. By breeding the Campbell crosses back again and again to pure Welsh Harlequins, he could stabilize the breed whilst avoiding the dangers of too much inbreeding. He called this flock his 'Harlequin Campbells'.

Most of the Welsh Harlequins available today are the product of Eddie Grayson's promotion and his formation of the Welsh Harlequin Duck Club, which managed to re-standardize the breed following an inaccurate version published by the Poultry Club in 1982. The new Standard, submitted first in 1986–7 to the British Waterfowl Association, was eventually included in the *British Poultry Standards* of 1997. Like the Abacot Ranger, they produce over 200 hundred white eggs a year and seem to have found favour in the USA, arriving in 1968. They are also standardized in Germany and Australia.

GENERAL-PURPOSE BREEDS

ORPINGTON

fact file	
Classification	Light
Weight range (male)	2.2–3.4kg (5–7½lb)
Weight range (female)	2.2–3.2kg (5–7lb)
Utility	Eggs and table
Colour	Buff
Origin	UK
Flight potential	No
Status	Rare (UK), threatened (US)
Eggs	Over 200

The first record of Orpington ducks (as opposed to Orpington chickens) we could find dates back to March 1898, when William Cook of Orpington was advertising the sale of ducks and eggs from his Buff and Blue varieties. This contradicts information in many texts, particularly most of the Poultry Standards in the 20th century. It supports the admission by Mrs Campbell that she was trying to emulate Cook's success with buff ducks (see page 81).

Prior to the creation of the Orpington, Cook had already been experimenting with crossbreeds. His 1894 book *Ducks: and how to make them pay* identified a number of interesting crosses: Rouen–Aylesbury, Pekin–Rouen, Pekin–Aylesbury, Muscovy–Aylesbury, mallard–Rouen. He was also a fan of the Indian Runner. It is not surprising, therefore, that he should have tried to blend a number of bloodlines (Runner, Cayuga, Aylesbury, Rouen and later Pekin). The potential colour forms from this would be kaleidoscopic, especially in later generations. The trick was to breed enough ducks to be able to choose the interesting variants, then he had to mate 'like with like' and end up with stable colour forms by trial and error. Modern genetics would have cut down the process considerably, but Cook was a pioneer and the results were spectacular.

By far the most successful Orpington duck was the famous 'Buff'. Named after its plumage colour, with perhaps a smile at the East Kent Regiment (the 'Buffs') with their buff uniform facings, this variety was the perfect compromise between the Indian Runner egg-layers and the body muscle of the Aylesbury and Rouens. It was an excellent general-purpose duck, cheaper to keep and hardier than the heavyweights, and more useful for eggs on the farm and the smallholding. Glistening gold in the sunlight, it also looks something special – the perfect cover picture for a countryside magazine.

Unfortunately the Buff Orpington does not breed true. This was identified by breed authority A F M Stevenson in 1926. He saw three duckling colour forms from the same buff parents: soft olive yellow down with yellow horn bill; pale yellow down with yellow bill; brownish shade of down with dark brown bill and legs. It took until 1963 for the genetics of this to be fully uncovered. F M Lancaster explained it again in his 1980 article 'Revitalizing the Buff Orpington'.

In the adult plumage of the Orpington these three forms comprise:

Buff Intermediate head colour in the males (seal brown), a dark grey brown, with very slight indication of blue on the rump. Both sexes have an overall even buff body colour with little evidence of pencilling.

Blond Males pale buff with a light grey brown head and possibly more blue on the rump. Females paler than the Buffs, with even less chance of pencilling.

Brown Light khaki pencilled plumage on the females; brown heads and rumps on the males; no evidence of blue.

In the USA, the breed is known as 'Buff'. It was imported in 1908 and admitted to the American Standard of Perfection in 1914. Buff Orpingtons have also become popular in Germany and Australia.

CRESTED

fact file

Classification	Light
Weight (male)	3.2kg (7lb)
Weight (female)	2.7kg (6lb)
Utility	General purpose
Colour	Various
Origin	Unknown, possibly Holland
Flight potential	No
Status	Critical (US)
Eggs	150–200

There are Dutch paintings dating back to the 17th century showing ducks with what look like pompoms on their heads. Charles Darwin also records 'tufted' ducks from the Far East. Here is another example of a breed that may have been imported into Europe by the Dutch East India Company, leaving visual evidence but little in the way of written material. White Crested ducks were standardized early in the USA, in 1874, and had their own club in the UK in the early 20th century. Standardized in the UK in1910, they are now exhibited in any colour.

Crested breeds are not for beginners. They are highly ornamental, but their appeal is reduced when one realizes how the crest is formed. It is the result of what is known as a 'lethal gene'. In its impure form, this produces a hole in the top of the skull (a cranial hernia) through which the fatty tissue covering the brain protrudes, causing the feathers on the crown to mass into a tufted crest. In the pure form, most infants are badly deformed or 'dead in shell'. It is therefore not advisable to mate two crested birds. Judges always need to check the spines of Crested Ducks: many show spinal deformities in the shape of kinked necks, roach backs and twisted tails. They may also suffer from a lack of co-ordination. Fortunately, the effects of the gene are not always expressed: some ducklings have no crest at all; others have very reduced ones.

In the show pen, Crested Ducks are judged almost entirely on the quality of their crests. A large, globular crest is the aim, and it should be 'simple' (not having a multiple tuft), central on the crown, not slipped to one side, nor too far to the back or impeding the eyes.

Crested are kept mainly as pets, but they are also quite good layers.

MAGPIE

fact file	
Classification	Light
Weight range (male)	2.5–3.2kg (5½–7 lb)
Weight range (female)	2–2.7kg (4½–6lb)
Utility	General purpose
Colour	Coloured and white
Origin	UK
Flight potential	No
Status	Rare, (UK), critical (US)
Eggs	150–200

Black ducks, Fawn-and-white Runners and ducks with bibs all provide the genetic material for the Magpie in its original pattern of black and white. If you imagine a white duck with a black 'cap', black back, rump and tail, and a mantle of black shoulder feathers (covering the white wings when the duck is resting), that is the simple Magpie pattern.

It was developed as a hardy, free-ranging commercial bird. However, it did not come out top in 1920s laying trials; it is at the 'heavy' end of the light ducks. It was a popular exhibition bird in 1920s and '30s, and advocated by Reginald Appleyard as a dual-purpose breed. It then almost disappeared until its revival in the 1980s. The breed was standardized in the UK in 1926, and in the USA in 1977.

Developed in Wales shortly after the First World War by Reverend Gower Williams and Mr Oliver Drake, this breed has an equivalent in the German Altrheiner-Elsterenten, produced later by Paul-Erwin Oswald.

Getting this breed show-perfect is the real challenge. Black spots in the wrong places or asymmetrical markings conspire to ruin the exhibition specimen. Out of a batch of several dozen ducklings, very few if any will end up show-perfect. There is a lot of wastage, unless the adults are sold as pets or table birds. As a result, if you are not too fussy about markings, these birds are available early, even as ducklings, as good layers and pets.

Blue and Chocolate Magpies have also been developed.

BANTAM DUCKS

Bantams are a third to a quarter the weight of full-sized ducks. Attractive, tame and economical to keep as pets, they can also lay a reasonable number of eggs (around 80), which is more than most Calls. As a bonus, the eggs are larger and so better for cooking.

Bantam forms of the Saxony and Welsh Harlequin are popular in Australia. Innovations in the USA are the Blue Indie and the 'Australian' Spotted, which has nothing to do with Australia, just as the Indie has nothing to do with the East Indies!

Note that all Bantams need clipping (see page 37). They make good pets and are useful 'broodies'.

SILVER BANTAM

fact file	
Classification	Bantam
Weight (male)	0.9kg (2lb)
Weight (female)	0.8kg (1¾lb)
Utility	Pet
Colour	Silver
Origin	UK
Flight potential	Yes
Status	Not widely kept UK
Eggs	30–80

Popular in the 1980s, the Silver (Appleyard) Bantam has lost out recently to the Silver Appleyard Miniature, which more accurately mimics the plumage of the large version. The Bantam has a colour form similar to both the Abacot Ranger and the Silver Call, now a favourite of the showing fraternity.

According to John Hall, the Silver Bantam was created in the 1940s by Reginald Appleyard from a cross between a small Khaki Campbell female and a White Call drake. The illustrations in Jean Delacour's 1964 edition of *The Waterfowl of the World* show a pair of 'Silver Bantam Ducks': the male very similar in markings to the modern specimen; the female more like a Silver Appleyard Miniature, with more extensive brown pencilled plumage and eye stripes.

SILVER APPLEYARD MINIATURE

This is another breed developed or created by Tom Bartlett (see page 79). Tom explained that he 'bred down' specimens of the large Silver Appleyard, but the way he did this was never fully documented.

Early photographs in *Fancy Fowl* magazine show beautiful examples of female Appleyards ('like porcelain figures'), almost perfect in markings. The drake had a very broad neck ring, suggestive of possible Pied Call connections. Nevertheless, Tom produced the Miniature in time for the first BWA Champion Waterfowl Exhibition held at Malvern in 1987. It was standardized ten years later.

Small, economical, tame and excellent at brooding difficult Call duck eggs, the 'Mini App' is an attractive and useful addition to the garden flock. It is also a great choice for a beginner in duck breeding, or an expert looking for a challenge. Keeping them pure and not getting them too dark are the prime targets for the responsible breeder.

Minis lay surprisingly large eggs. Keep taking them away from the nest if you don't want the bird to sit. As with Calls, the birds must be confined to the garden in the breeding season; they do not lay before 9 a.m. and a broody duck can be lost on her concealed nest.

fact file	
Classification	Bantam
Weight (male)	1.4kg (3lb)
Weight (female)	1.1kg (2½lb)
Utility	Pet
Colour	Restricted Mallard
Origin	UK
Flight potential	Yes
Status	Popular breed UK
Eggs	80

CALL

fact file	
Classification	Call
Weight range (male)	0.6–0.7kg (1⅓–1 ½lb)
Weight range (female)	0.5–0.6kg (1–1 ⅓lb)
Utility	Pet
Colour	Various
Origin	Probably Holland
Flight potential	Yes
Status	Popular breed UK
Eggs	0–30

Call Ducks are small, cute, cuddly and quite noisy – at least the females are. For those not experienced with ducks, it is only the females that quack. The males are reduced to a lisping squawk.

It is a dwarf gene that reduces the length of the bill and body, and produces a large, round head, chubby cheeks and round, bath-duck body. In spite of the size, this breed has 'attitude'. Call ducks show little respect for and no fear of humans. A drake protecting his female's nest will squawk and head-butt your wellies. A hungry female, waiting to be fed, will get under your feet and demand attention. There is a real danger that they will be trodden on. However, they are great pets.

Originally named 'decoy' ducks, then 'call' ducks, these little birds were selected to entrap mallard, teal, widgeon and other wild ducks. Their tolerance of humans and their loud 'call' encouraged the wild birds to come within range of the 'decoys' (from the Dutch for traps), large spider-like contraptions that funnelled the intruding ducks deeper and deeper until there was no escape. The early decoy ducks, though, were little different from the wild mallard, being decoys by training and performance rather than looks. Dutch Call ducks were the end result of mutation and selective breeding, possibly reinforced by imports from the Far East at the end of the 18th century.

In 1848 the Reverend Edmund Saul Dixon wrote about a 'smaller race of White Ducks imported from Holland'. Other writers ascribed the little breed (known as mignon) to France or Italy. Few ducks can claim an accurate written pedigree, yet by 1865 these were well enough established for a Standard to be drawn up, for just two colour varieties: white and grey (mallard-coloured). Then they dropped out of the Standards altogether for nearly a hundred years, confined to the second division of 'ornamental' waterfowl.

Nowadays they occupy a place at the very top of the premier league. More Calls are exhibited in the UK than any other breed of duck, followed closely by the Runners. More colour varieties are now standardized than for any other breed, again followed by the Runners. For birds with such short legs, one would have expected a different outcome to that particular race. The 2008 BWA Standard recognizes 17 varieties of Call ducks and 14 of Runners. It is getting close.

Exhibition Calls lay few eggs and hatch even fewer. A combination of inbreeding and extremely small eggs leaves little margin for error. Bantam duck broodies are more effective than most incubators (or a combination of the two). Broody bantam chickens can be good, as long as they do not puncture the eggs. Humidity can be a critical factor, as the eggs have a tendency to dry too slowly, or even too fast in some climates. Little Call eggs are not as reliable as Runner or light duck eggs, whose ducklings seem almost to bounce out of the shells. Call ducklings should be kept in coops much longer than bigger ducks. They are prime targets for crows, magpies, ravens and cats and need lots of care.

So too do adult Calls. The white ones in particular can be targets for predators. They can also fly, so you may need to trim the feathers of one wing (see page 37) or keep them safely under nets. Keep an eye on them at breeding time: wandering off and nesting through the hedge and then disappearing altogether is a favourite trick. You can guarantee that a fox will find them before you do. Lastly if a duck is looking seedy and sluggish in the laying season, she may have a trapped egg or one without a proper shell. The little pelvis may be too small for her to lay easily. Seek the help of your vet, but do it quickly.

BLACK EAST INDIAN/BLACK EAST INDIE

fact file	
Classification	Bantam
Weight range (male)	0.9kg (2lb)
Weight range (female)	0.7–0.8kg (1 ½–1 ¾lb)
Utility	Pet
Colour	Black
Origin	Probably USA
Flight potential	Yes
Status	Popular breed UK and US
Eggs	30–80

The Black East Indian is just a shade larger than the Call Duck, but it has a very different body and head shape and spectacular plumage: deep black with strong green reflections. The Germans call it the 'Emerald' duck. Slightly smaller than the wild Mallard, it looks like a diminutive Cayuga. As Paul Ives put it, 'In 1943 the committee of three professional artists invited to select the most beautiful bird in the Boston Poultry Show…selected a Black East Indian drake as the most beautiful bird among 5000 specimens…'

The bill should be as black as possible, with little sign of grey and no light green. The show fraternity also frowns upon purple highlights. Green should dominate over as much of the plumage surface as possible. Shape and size also play significant parts in the judging decision.

These birds make great little pets. A flock of Black East Indians on the green grass in summer looks super, the sunlight shining off their lustrous black feathers – black they may be, but dull they are not.

Where they come from is the real problem. Black East Indies? Not much of a chance. Brazil? Well, they were called 'Brazilian' or 'Buenos Ayres' at one time, but there is no evidence for this being the origin. They were also known as the 'Labrador' duck, and they may have evolved in North America, like the Cayuga. Nonetheless, we still call it the Black East Indian, which is just as appropriate as naming a bird that comes from China the African goose!

The Black East Indian duck brought black genes to Britain some time in the early 19th century. Documentary evidence suggests that it came into the possession of the London Zoological Society when Lord Derby was elected President in 1831. The breed was first standardized, along with the Call Duck, Aylesbury and Rouen, in 1865. Like the Cayuga, early eggs are coated in black, but change to white (or green). Number of eggs and hatchability depends upon the strain.

All About Eggs

Eggs for eating, hatching, cooking or decorating are great fun to produce. But they don't just happen – the birds must be the right age to lay. Look after females carefully in the laying season: they are under much more stress than males. And if you want to enjoy breeding from them, choose parents carefully for best results.

Egg production

Ducks vary tremendously in their egg-laying capacity. Small Call ducks may lay only a few eggs per year, though some strains manage between 30 and 80. The heavier breeds often lay 120–150 eggs per year, while the light breeds like Welsh Harlequin and Khaki Campbell produce far more. Commercial strains of Campbell often achieve more than 300.

Ducks must be mature to start to lay. That means a minimum of 21 weeks of age in the Campbells; 26 weeks is more likely. The Rouen and the true Aylesbury lay in the traditional breeding season, spring; in contrast, Campbells will lay most of the year, given the right lighting conditions and correct food. Up to 15 hours of light per day (natural or artificial) is recommended for best production, but ducks can still lay in December, even with no additional light. Age is the more important factor: ducks lay best in their first two years, but the number of eggs really falls off after year four. Birds which have relatively low production will carry on laying for longer: Calls are known to lay a few eggs even at ten years of age.

Geese are much more seasonal creatures. Most breeds do not start to lay until they are almost a year old. Their eggs can be small in their first year, and they become better breeders from year two onwards. Females can continue to lay even after the age of 20, but their best reproductive years are between two and 12. Average-sized breeds of geese have a longer reproductive life than Chinese and heavy geese

A few birds will lay eggs in the autumn. This is usually associated with artificial light stimulating them, but young Chinese can lay in October. The number of eggs varies very much according to the breed and feeding/management.

No eggs?

Ducks need time to moult, and when this happens they stop laying. Growing new feathers and laying eggs make huge demands on protein and nutrients, and the birds cannot be expected to do both at the same time. Both ducks and geese also stop laying when they feel broody.

If birds should be in lay but there is no sign of eggs, check for possible hiding places. Eggs may be buried in the litter of the shed, or concealed under hedges. In all but the smaller breeds, ducks generally lay before 8 a.m. So if the eggs are being dropped in the field and the crows are taking them, or if they are being popped in the pool, let the ducks out a bit later in the morning.

The laying cycle of the goose is around 36 hours. So eggs can be expected every other day. Geese are also adept at hiding their eggs in litter, making nests under hedges – or dropping eggs in the field for crows to carry away. Check what is happening if there appear to be no eggs. Occasionally, birds simply do not lay.

Birds do sometimes eat their own eggs. A duck or goose rarely deliberately breaks one to eat the shell. More commonly, any eggs eaten were first broken accidentally.

Production will of course be low or stop if the birds are ill, underfed, have poor-quality food or are short of water.

The ovary and oviduct

In nearly all birds, only the left ovary develops. It contains a large numbers of oocytes (the 'germ cells' that give rise to eggs). When an oocyte is preparing for ovulation, it develops yolk materials around it. The ovary will contain several oocytes at different stages.

At ovulation, an oocyte passes into the first part of the oviduct, called the infundibulum. Occasionally, two oocytes are released at the same time. This will result in a double-yolked egg. Fertilization occurs at this point, and it is possible to have a fertile, double-yolk egg. It is best not to incubate these, of course.

As the egg passes down the oviduct, albumen (which is mainly protein, minerals and water) is added in the magnum, and the spiral chalazae, two rope-like cords which anchor the yolk into a central position, also develop.

After the membrane around the albumen has been added in the isthmus region, the egg reaches the shell gland and the outer layer is calcified. The shell is thin but very strong. Colour (which varies with the breed of duck) is also added to the outer shell at this stage. The egg then passes down to the cloaca and is laid, the whole process taking 24 hours or more, depending on the species. The warm, moist egg has a coating which quickly dries to form an anti-bacterial seal, closing the pores in the eggshell.

HOW EGGS ARE MADE AND HOW FERTILE EGGS GROW

The diagrams below show the process of the chick developing within the eggs. If all goes well, a healthy duckling or gosling will emerge within 4–5 weeks, depending on the species (see box on page 100 for more details).

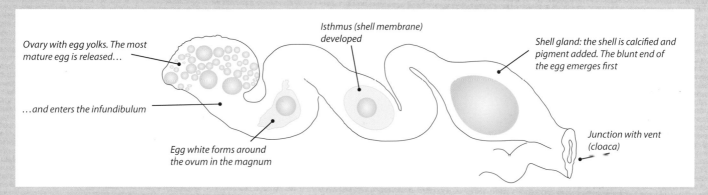

Ovary with egg yolks. The most mature egg is released…

…and enters the infundibulum

Egg white forms around the ovum in the magnum

Isthmus (shell membrane) developed

Shell gland: the shell is calcified and pigment added. The blunt end of the egg emerges first

Junction with vent (cloaca)

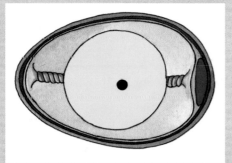

DAY 1: There is no development until the egg is incubated but the germinal disc can be seen as a small spot.

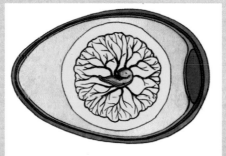

DAY 5: A network of veins grows and the heart is formed.

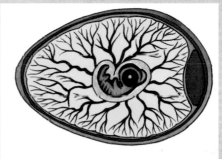

DAY 8: The head and the eye are well developed, and the heart is enclosed within the body.

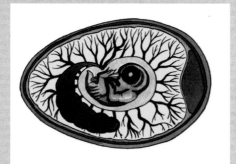

DAY 12: The waste products are collected in a sac which is in fact developed by day 4. Feet, wings and beak are formed.

DAY 27: The air sac has grown because water has been lost.

HATCHING: The duckling breaks into the air sac with its egg tooth, then cuts its way out of the shell.

Egg quality and structure

Both the quality of the shell and the internal egg structure and nutrients are influenced by the diet and lifestyle of the bird. Since the egg contains all the nutrients essential for the production of a duckling or gosling, parent birds should be fed the best diet possible. Eggs will not start to develop unless (a) they are fertilized and (b) they are incubated. If eggs are taken away each day, the birds are less likely to go broody and will continue to lay for longer.

The shell surface is penetrated by pores which go through the testa to allow the exchange of gases with the shell's interior – essential during incubation. The growing embryo uses some of the calcium carbonate in the shell to form its bony structure as it grows.

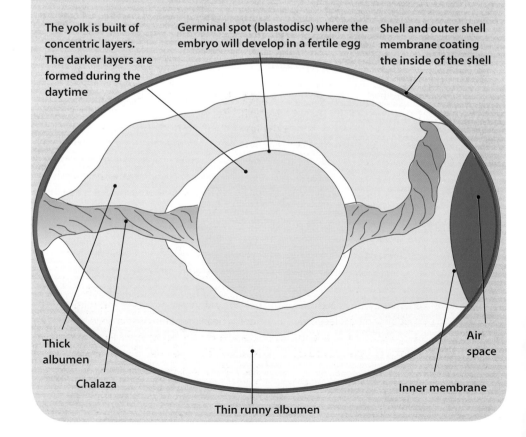

INSIDE THE EGG

The chalazae are ropy structures which hold the yolk in the centre of the albumen. They coil in opposite directions and allow the yolk to rotate so that the germinal spot will float to the top when an egg is turned (see page 100). This is so that the embryo is always in contact with the warmth of the sitting bird. Eggs must be turned to allow the embryo access to fresh nutrients.

During incubation, the embryo uses the nutrients in the albumen, and the air space (air sac) expands as water is lost through the pores. The larger air sac allows the hatchling to breathe when it breaks through the inner membrane, and have enough room to manoeuvre to get out of the shell.

The yolk is built of concentric layers. The darker layers are formed during the daytime

Germinal spot (blastodisc) where the embryo will develop in a fertile egg

Shell and outer shell membrane coating the inside of the shell

Thick albumen

Chalaza

Thin runny albumen

Air space

Inner membrane

EGG SIZE
Ducks eggs vary immensely in size, depending on the breed. The three smallest eggs above are Call at 40g (1½oz) and 46g (1¾oz) and Miniature Appleyard at 55g (2oz). The green and white Runner eggs (left) are around 70g (2½oz).

Cayuga ducks can lay almost black eggs at the beginning of the egg-laying season. This coloured cuticle gradually becomes less pronounced as the season progresses. The two white goose eggs are 160 and 212g (5½ and 7½ oz). Geese always lay white eggs.

Eggs for eating and selling

Collect all eggs daily and store at around 12⁰C (54⁰F). Hen's eggs (clean and unwashed) can be stored for as much as four weeks, as can good-quality goose eggs. However, all eggs are better eaten within two weeks.

Duck eggs are said to be more porous than hen eggs, and they often need washing. Ducks and geese lay on the floor of the house: provide a low, boxed-off area containing chopped straw and/or clean wood shavings for them to do so. The provision of clean litter avoids contamination by muddy or mucky feet and droppings. If eggs are soiled, or laid outdoors and muddy, they should be washed and used within ten days.

Fewer regulations apply to the sale of duck and goose eggs than chicken eggs. This is because they are not a significant part of the food chain. They are usually sold locally, in small quantities, and are therefore traceable because they are 'farm-gate' or delivered by hand. They need not be graded for size or quality, or stamped with a producer registration number.

Do ensure, however, that the shells are of good quality and unbroken, because then the eggs will not admit bacteria so easily and will store for longer. Cracked but fresh eggs can be used at home if well cooked e.g. in cakes.

Egg quality

Duck eggs have long had a reputation for carrying salmonella. However, if the birds are kept in clean conditions with a good water supply, and if the eggs are cooked correctly, this is not a risk. If anyone is wondering how long to boil a goose egg for, the recommended time is 9–11 minutes.

It was once thought that the cholesterol in eggs was bad for human health but the reverse is now thought to be true. Eggs are a good source of high-quality protein, and egg yolk is one of the best sources of the beneficial nutrients lutein and xenazanthin. A good supply of lutein and xenazanthin is thought to help prevent age-related macular degeneration, which is a leading cause of vision loss amongst people over 65. Lutein also appears to inhibit the development of atherosclerosis.

Birds fed a free-range diet high in greens also produce healthier eggs for humans to eat. The egg supplies complete protein and vitamins A, B and D. Almost half the fat in eggs is healthy monounsaturated, the kind found in olive oil.

Eggs should not be incubated or consumed if the ducks have been on medication. Antibiotics and wormers have a withdrawal time indicated on the product label. Organic products such as Vermex do not need a withdrawal period.

Breeding Ducks and Geese

Breeding animals and birds can be a really satisfying hobby. However, it is not an experiment to be undertaken lightly and it might be helpful to run through this check list first.

1. **Decide what will happen to the birds before the eggs are even set (put in an incubator).** If the birds are for your own use/consumption, there it no problem. If they are to be sold, then costs in advertising, transport and marketing should be taken into consideration.

2. **Don't hatch more than you can cope with.** This is especially important with waterfowl. They need more space and water than chicks. Ducks make an awful mess unless they have well-designed facilities; goslings need clean grass. Set up your rearing equipment before breeding.

3. **Set eggs in the natural breeding season, from good-quality parents.** Geese and most ducks limit their laying season from late February until June, generally with a break in the middle. Eggs set during the earlier and middle part of that time will produce the best birds.

4. **Learn how to incubate eggs by broody bird or electric incubator by reading reference material first.** Too many people embark on hatching a few eggs and only begin to find out what to do when the process has already started.

5. **Decide if you want to use natural incubation or incubators.** Incubators are expensive if you end up using them only once. Birds can do a better job, especially with eggs which are known to be 'hard to hatch'. But managing broody birds is a whole topic in itself, and a back-up incubator (to use as a hatcher) is useful for an emergency. Natural rearing needs as much forethought and planning as artificial incubation.

RIGHT *Muscovy ducks are good sitters and mothers. However, free-range ducklings are easy prey for magpies, herons and pike on large pools and canals, so be cautious about the available water.*

Eggs for hatching

Eggs for hatching should be produced from fit birds kept in good conditions, preferably fed on breeder pellets, and on free range or with garden access. The parent birds should not be too closely related: hatching eggs from inbred birds can lead to genetic problems such as spinal deformity and bad feet and legs. Also, eggs from inbred birds are commonly infertile.

Parent birds must be separated from a group of mixed birds for at least 14 days to guarantee the father. If the duck has been with totally unrelated drakes, increase this to 18 days.

If eggs are dirty, don't use them. Germs can penetrate a dirty, wet shell. There is no point in growing bacterial colonies inside eggs which will then go bad and explode in the incubator. Give birds nesting places with plenty of clean litter such as wood shavings and chopped straw. It is better to preserve the outer layer of the egg (the cuticle) intact. This is especially important for natural hatching because the cuticle is a protective layer designed to keep out harmful bacteria. Washing removes this layer. If the eggs must be washed, use a nail brush with running, warm water (hand hot, around 40°C/104°F). This will prevent the egg contracting and drawing water and harmful bacteria through the shell pores. Egg sanitizer powder can be obtained from specialist companies such as Interhatch. For smaller batches of clean eggs in a clean incubator, sanitizer is not necessary.

Also, use only eggs with good-quality (not rough or thin) shells, and of normal size for the bird. Mark each egg in pencil with the date laid and its parent.

It is best to allow eggs to rest for at least one day and a maximum of ten days before setting them in an incubator; so from one duck you could have a clutch of up to nine eggs. Collect and clean eggs daily and store them in a cool room at around 12°C (54°F) so that they do not start to develop. If possible, store them vertically, with the blunt end uppermost; if you have to store them on their side, hand-turn them daily.

The night before you put them in the incubator, raise the temperature to 20–25°C (68–77°F) to prevent the yolk membrane rupturing due to expansion caused by a rapid rise in temperature.

Buying eggs for hatching

This is a risky business. There are lots of variables you are unable to check on, including the breed of the duck or goose: the vendor may be misinformed, or the eggs could be crossbreeds or infertiles. Call duck eggs are particularly difficult to hatch, and they are often infertile early in the season (January to March). In summary: don't pay a lot of money for eggs that you really know nothing about. Also beware of buying eggs from abroad. Without a licence, this would be illegal trade. The USA and Australia are particularly stringent on imports because of concerns about biosecurity (see page 110).

Aim to have a synchronized hatch. Fertile eggs do not develop until they are incubated, and it is far better for the birds to hatch together and have companions to keep each other warm and to imprint upon. As we have said, it is not kind to rear one bird which imprints just on its owner. Always have bird companions.

Incubating eggs must lose water

Essential to the hatching process is the loss of water from the egg as the embryo metabolizes its food. Water loss through the pores of the shell is often quoted at 13–14 per cent. Wild birds lose 16 per cent and 15–16 per cent is better for waterfowl. Water loss, largely from the albumen, allows the air sac (at the bulbous end) to enlarge, enabling the hatchling to break into the space and breathe in air just prior to hatching, and then giving it room to manoeuvre to get out. Eggs which lose insufficient water are often 'dead in shell' and have not drawn their yolk sac into their body.

There are a number of things that affect this delicate process of water loss:

- the **porosity of the shell**, which varies with breed, diet and age of the bird.
- the **relative humidity** (RH; that is, moisture) of the ambient atmosphere during incubation.
- the **size and type of incubator** used: these govern the passage of air over the shell. Still-air incubators need less water than the forced-air type where the fan, which maintains an even temperature, also aids evaporation because the air is moving. See the next page for more about this.

Incubators have manufacturer's recommendations for best results: for example, goose eggs should be incubated at RH 50–55 per cent. This may work for large incubators with 'easy to hatch' commercial strains, but not for small, still-air incubators with pure breeds. Note that waterfowl eggs do not need more water than chicken eggs; generally they need less. Correct water loss is crucial to successful hatching.

WELFARE WARNING

Newly hatched ducklings and goslings are the most appealing birds. Fluffy, talkative and quick on their feet, these precocial young are just one of the rewards of keeping waterfowl. However, before getting too enthusiastic about hatching them, you have to take their welfare into consideration. Looking after, rearing, maintaining and, even selling have to be taken into account. There is no point in breeding livestock that you cannot easily look after or sell. Only breed birds if you have good facilities and have planned what will happen to them: practise good welfare.

Artificial incubation: electric incubators

There are many different models of small incubators on the market. They are made in two basic types: still air and forced air (with a fan). Still-air models have only one layer of eggs. The forced-air models can have more than one layer and often have automatic turning. Some also have humidity control. Read the manufacturer's specifications carefully to ensure you buy something that suits your needs.

Eggs of Chinese geese are easier to hatch than those of other breeds. In ducks, Runner eggs hatch far more easily than Call duck eggs, which really benefit from natural hatching under broodies.

Temperature

Run a new incubator for a day first to check that it is working correctly i.e. the temperature is stable and the thermometer is correct: check it against a clinical thermometer. The recommended temperature for goose eggs in a forced-air cabinet varies between 37.2 and 37.7°C (99–100°F); a typical duck-egg temperature is 37.3°C (99.1°F). With a fan, the temperature is the same throughout the cabinet.

In a still-air incubator, the temperature is often measured by a thermometer on a stand above the eggs. But because hot air rises, the temperature on the floor of the incubator is lower and there may be a difference of as much as 2–3°C (4–5°F) between the floor and the top of a large egg. So the temperature for incubation is correct at only one level. The recommended temperature for goose eggs is around 39°C (102°F) at a point just above the eggs.

With all models in the Brinsea Octagon range, egg turning is achieved by rotating the whole cabinet through 90° (45° on either side on an hourly cycle). The separate turning cradle keeps all moving parts outside the incubator away from moisture and dust, leaving the enclosure easier to clean. It also means that there is nothing to trap or injure emerging chicks if you forget to stop the turning. The incubator lifts easily out of the cradle, allowing it to stand horizontal for hatching.

CANDLING EGGS: FERTILITY AND HUMIDITY

The most valuable aid for success in hatching is a good candling lamp, which shines a light through the egg in a dark room. It is very useful for checking fertility and monitoring the growth of the air sac as the egg loses water. However hi-tech your incubator, the numerous variables involved in water loss mean that observation is essential for success.

With experience, it is possible to tell visually if the eggs are too wet or too dry with a fair degree of accuracy. However, there can be a rapid change in the air sac at about day 22 when the embryo can 'drink' its own fluids – the sac may suddenly grow large and the embryo look very dry.

A fine spray of warm, clean water once a day is often recommended for eggs close to hatching, but this is only good practice if the air sac is much too large. If the air sac is ideal, leave well alone. Damp conditions allow bacteria to grow. Ducklings with correct moisture loss do not need excessive moisture for hatching: filling the water trays in the bottom of the incubator, when half the eggs have pipped, is ample. This should, of course, be done earlier if eggs are losing too much water.

To maintain high relative humidity for eggs which are dehydrating too rapidly, spread a wet cloth over the unused portion of the incubator (don't let it touch the eggs). It is the area of evaporating moisture which makes the biggest difference to RH. Note that evaporating water can initially drop the temperature of a small incubator; it is essential to monitor it. This strategy is also useful when half of the eggs have pipped.

Thicker-shelled goose eggs are more difficult to candle than duck eggs, and need a much stronger light. Eggs under broody birds should also be tested for fertility: there is no point in broodies sitting on infertile or rotten eggs. Rotten eggs can of course be identified by their smell.

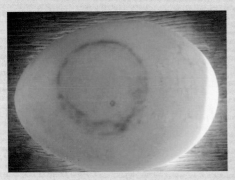

Day 8: This egg is dead: the veins have collapsed into a 'blood ring' at seven days.

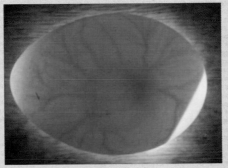

Day 8: This fertile egg shows a network of veins after eight days of incubation. The darker area is the head and eye. The air sac has already visibly grown.

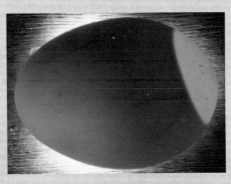

Day 12: The darker head and eye are more visible.

Day 24: Nearly the whole egg is dark.

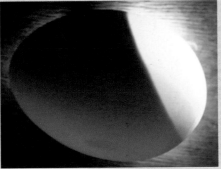

Day 27: The air sac has grown to an ideal size, and the duckling can be seen moving inside the membrane. It is not yet breathing air.

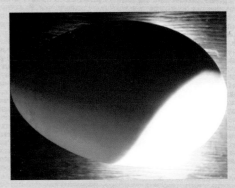

Day 28: This egg failed to hatch; it lost too much water and did not pip. The lighter rim inside the dark area suggests that the duckling was dead.

During incubation

- **Turning**: In an automatic-turn incubator, the eggs are placed on their side (or on their end) and turned each hour. In a hand-turn, still-air incubator, turn the eggs at least three times a day. Turn them in the opposite direction each time, otherwise the internal structure could be disrupted.
- **Recording**: In a manually operated incubator keep a recording sheet with the date of setting, expected date of hatch and a record of the turns (left then right). Of course, the automatic turners will do all that turning for you. Turning should be stopped around day 25, or the eggs removed to a still-air hatcher.
- **Candling**: Buying a candling lamp saves a lot of problems. It is important to candle eggs to check if they are fertile (at 7–8 days) and to see if they are developing well (at 14 days). At this stage, remove any eggs which are going bad; otherwise they may infect the whole incubator. Discard any cracked eggs too. Candling also enables you to monitor air sac development (see page 99).
- **Water and humidity**: The amount of water needed in the incubator depends on many variables. Even with a hygrometer and humidity control in advanced incubators, the manufacturer's figure will not fit every egg. The correct amount of water needed is something each operator has to learn for their individual circumstances. Egg porosity varies a great deal and in Britain environmental relative humidity varies enormously too.

INCUBATION PERIOD

Waterfowl eggs take longer to hatch than chickens. The incubation period varies with the species. If eggs go cold, perhaps because of a power cut or because a broody bird deserts the nest, incubate them anyway and see what happens. The eggs may still hatch, a little late.

Species	Days
Domestic goose	28–32
Domestic duck	28–30
Call duck	26–27
Muscovy	33–35

The exact time depends upon variables such as water weight loss, age of the egg when set, and viability of the embryo. Fresh eggs from unrelated birds, with correct weight loss, hatch earlier than those with problems. However, expect a hatch to be complete over a 2–3 day period.

LEFT *During incubation, these eggs have been automatically turned by the moving floor in this Brinsea Polyhatch. The turning mechanism is switched off 2–3 days before pipping, when ducklings are moving themselves into a suitable position for hatching. Humidity must be raised when the eggs are pipped (around day 28 on average for ducks) so that the ducklings do not get glued in their shell. Keep ducklings in the incubator until they are dry and fluffy.*

RIGHT *Goslings are slower at getting on their feet than ducklings, but are then much more aware of their surroundings and keepers. Hand-reared, they will stay very tame. Very clean straw bedding may be chewed but cannot be eaten in mistake for food.*

Hatching: don't interfere

When the hatchling first pips the shell, it is not ready to hatch. Usually, it needs at least one or two days to complete hatching. The yolk sac must be drawn inside the body and blood from the veins absorbed first. If you interfere, it will bleed. Only hatch a duckling/gosling which has nearly completed a full turn around the shell and if it seems to be stuck. Occasionally, the end of the egg shell is not flipped off, and you will need to help the bird. Birds which fail to hatch themselves often have something wrong with them. Occasionally birds pip at the wrong end of the shell. They take longer to hatch (after pipping) than normally positioned birds and you may eventually need to help them.

Although incubator manufacturers may say that their cabinet incubators can both incubate and hatch birds, this causes problems. Moving air dries out eggs as they pip and glues them up. When they do hatch, the hatching fluff gets blasted round the whole incubator, is difficult to clean up and becomes a bacterial hazard.

- Keep a forced-air incubator clean, and do not use it for hatching.
- Hatch in a still-air incubator. The humidity can be raised much more effectively in a still-air hatcher – and the main incubator also kept at the correct humidity for eggs which continue to incubate. A small incubator like a Brinsea Hatchmaker can be easily cleaned and sterilized after each hatch.
- If you have one incubator: set, incubate and hatch only one batch before sterilizing and starting again. Do not mix incubating and hatching.

It is very important to monitor the temperature in the last week and especially at hatching. The metabolic heat of the bird, and exertion at hatching, can raise the temperature of the eggs and a small hatcher to lethal levels. That's why it is important to be around at hatching time – to check temperature and humidity.

ABOVE *These Miniature Appleyard ducklings have still got their egg-tooth (at the end of the bill) which allowed them to cut their way out of the membrane and shell.*

BELOW *The eggshell on the left is pipped and the duckling is moving around to complete the full turn. The shell in the middle has been pipped at the wrong (pointed) end and the duckling will need help in getting out. The membrane is becoming dry, brown and leathery. The egg on the right looks dark, with too much moisture; it did not hatch.*

CARING FOR YOUR HATCHLINGS

Don't leave all the jobs of incubation and rearing to children unsupervised. You have a 'duty of care' to look after pet animals. Owners are required to provide a suitable environment for captive animals; adequate food and water; protection from pain, injury and disease; and conditions that allow animals to exhibit normal behaviour.

Natural hatching

Low-tech natural hatching can be more satisfying that hatching in the incubator. However, it requires a lot of observation and also the most work in keeping the conditions right.

- Firstly, the goose or duck must feel like going broody, though a broody hen can be used. A bird will not go broody until she has laid a clutch of eggs. Some breeds are more likely to go broody that others.
- The nest must be safe from predators.
- The nest litter should be clean. A turf, with a hollow beaten out, can be placed below the litter, to form a comfy shape and supply some moisture.
- A duck or goose will not accept being moved from her chosen place.
- The broody must be kept parasite-free. An organic dust such as Barrier is ideal, and should be used regularly. Don't use toxic pesticides just before hatching.
- Look after the sitting bird: she should be encouraged off the nest to eat and drink once or twice a day (depending on behaviour).
- Most important of all (especially for geese), worm the goose and her gander. Geese sit for 32 days, get very run down and are a frequent casualty in the breeding season if not well looked after.

On pipping, it is often best to hatch the eggs in the incubator, especially with Calls and goslings. Calls are very fragile; they are OK under a Call duck but often not with a broody hen. Geese can squash their goslings. An experienced goose will happily receive her goslings back, but a goose who has not hatched birds before must identify them in her nest first. Allow her to hatch just one viable egg, padding out the nest with 3–5 dummy eggs (hard-boiled eggs or those which look reluctant to hatch). Ganders are generally good parents (but do check; keep them away at hatching time and watch how they behave when the goslings are on their feet). Drakes are not trustworthy – remove them.

RIGHT *When a goose has laid a clutch of eggs and is ready to sit, she goes broody. She will then line her nest with down and feathers plucked from her breast and underparts.*

Welfare

Welfare is about spending time with the birds –
talking to them and observing them, as well as
cleaning, feeding and watering them. Nowhere
is that more important than in rearing young
birds. Waterfowl are great conversationalists
and time spent with them in the first few weeks
is well spent: it will determine their behaviour
towards humans for life.

Rearing

If you have not reared ducklings or goslings before, be prepared. They can create an awful lot of work. It is essential to find a way of supplying plenty of fresh water, and of maintaining hygiene, ensuring that the birds and bedding stay clean and dry because moulds and aspergillosis (see page 111) can develop in wet bedding.

Although ducklings and goslings can use the same starter crumbs, geese need more greens than ducks, both for nutritional reasons and to satisfy their natural instinct to graze. So be prepared to collect green material – grass and dandelions – to supplement their diet.

Ducklings and goslings need different approaches to floor type and water containment, so should ideally be treated as two separate groups even on a small scale, and definitely if you have a large number.

It's hatched!

Newly hatched waterfowl do not need feeding straight away. The yolk sac, which has been drawn into the body during hatching, will supply enough food and water for the next 24 hours, and the hatchling can survive up to 48 hours without food. However, in an artificial rearing environment, the drying effect of the heat source does require that drinking water be available, and that the ducklings and goslings are directed so that they know where it is.

As the birds grow, their food and drink arrangements need to change according to their size and demands. At all stages, great care should be taken to protect them from infections and predation.

FOOD

Duck and goose starter crumbs should be bought in advance; it may be difficult to get the correct rations at short notice. Standard chick crumbs have been known to kill ducklings such as Calls and Black East Indians because of additives such as coccidiostats.

Starter crumbs are designed for small mouths. They also contain around 18 per cent protein. As the birds grow, gradually replace the crumbs with duck/goose grower pellets, whose protein content is lower, at 15 per cent. Aim to replace all the crumbs with pellets by the time the youngsters are three weeks old, unless they are particularly small, like Call ducks.

Never allow goslings and ducklings to eat without water available. The food contains only 13 per cent moisture.

HEAT

The heat source must be reliable. Glass bulbs (red or clear) are not recommended. Dull emitters (ceramic bulbs) are more expensive, but last longer. They also allow natural light cycles for the birds, accustom them to darkness, and are less likely to malfunction

Over-heating will kill birds. The temperature under the heat source should be 35°C (95°F), falling to 32°C (90°C) during the first seven days, with no draughts, but the birds must have space to get away from the heated area. Monitor their behaviour: a tight group huddled under the heat source is cold; if they are dispersed well away from it they are too hot. Raise the height of the heat source in stages as the birds grow so that by the time they are three weeks old they are 'hardened off' to a temperature of around 20°C (68°F).

Water and surfaces

For the first 24 hours, keep ducklings and goslings on old cloth (no torn edges to chew) or crumpled, textured paper. Avoid smooth surfaces, which cause spraddled legs, especially in goslings.

Next, move goslings onto straw if possible. This avoids their eating shavings. They can move on to shavings for bedding when they know that it is better to eat crumbs, grass clippings and dandelions. There is less risk of moulds and bacteria developing in wood-waste bedding than in straw.

Goslings must be fed some green food, otherwise they will 'graze' each other. If reared with a goose, they must also be protected from predators. Even with a goose, they are best kept in a protected environment for three weeks. Magpies will still take fluffy goslings.

Ducklings will consume more water than goslings. Once they are on their feet, transfer them to wire-mesh platforms where they can use plastic 'drinking fountains', which are much easier to clean and sterilize than metal ones. Start with smaller 'drinkers' with a narrow rim, to prevent the risk of drowning. The ducklings can spend hours siphoning water out of the rim and transferring it to the floor. On a solid floor they will get wet, cold and thoroughly miserable. In their natural environment, water-proofed mallard ducklings are living on water by this stage.

ABOVE *These Runner ducklings are still fluffy at three weeks old. They can spend a few hours a day in a protected grass run if the weather is good.*

LEFT *Ducklings can be kept on a raised weldmesh platform for 3–4 weeks as long as they also have a mat. Newspaper works well: it should be renewed regularly and can then be composted. This system keeps the ducklings clean and dry, as long as you collect and remove the spillage underneath regularly. Do not let moulds develop on spilled food.*

Moving outdoors

Ducklings and goslings can be reared in protected grass runs once they are off-heat. This is likely to be when they are about three weeks old; exact timing will depend on the weather. Use a polythene cover in cool, wet weather. With goslings, you have to move the coop and run a lot more often than with ducklings, to enable them to eat clean grass.

If space is limited, or the weather cold and wet, ducklings can be kept on weld-mesh platforms off-heat for longer – but watch the condition of their feet. They must have plenty of water and an area of matting. Goslings cannot be reared like this. They need to graze.

Goslings, as well as adult geese, must have coarse sand available. The sand they eat punctures the grass blades in the gizzard to obtain the juices more effectively. Goslings should be grazed on clean ground, free from droppings. They are particularly vulnerable to gizzard worm, which can be passed on by adult geese. Adults that accompany goslings must be wormed. If you suspect that goslings may be suffering from worms e.g. if they are losing weight, add Flubenvet (see page 115) to their feed for a week. A worming drench prescribed by the vet can be used in an emergency.

Goslings must also be protected from excessive rain because the covering of feathers on their back is not complete until around six weeks. They are prone to pneumonia if cold and damp. Rearing them with a goose who broods them avoids this: she will keep them dry and well oiled.

Throughout rearing, birds must have the correct food. They also need to range on grass to eat, exercise and keep their feet in good condition.

RIGHT *The coop and run must be moved often for goslings to enable them to eat clean grass.*

Diet changes as the birds grow up

Starter crumbs and grower pellets are baby food: as the birds grow they need to eat whole grains as well. Start by mixing wheat in with the grower pellets, so that by the time the young birds are six weeks old wheat makes up 50 per cent of their diet. Feed this dry mixture in a heavy bowl or trough. Also offer – under water – small amounts of whole wheat, which does not disintegrate in water.

Goslings can proceed on to just grass and wheat as their basic diet after they have refeathered at 16–18 weeks. Female ducks which might come into lay at 24 weeks (early hatched Runners and Campbells) should be fed wheat and layer pellets from 16 weeks to build up calcium reserves for eggshell formation.

ABOVE *Over-feeding on compound food can cause rapid growth of the flight feathers. Then the 'blood quills', from which the flights emerge, become too heavy for the wings and turn outwards, causing 'angel wing' or slipped wing. This condition does not occur in wild birds. These blood quills are normal at five weeks.*

If the blood quills are not set correctly, lightly set them in the correct position with Sellotape. Remove the tape after three days. Repeat if necessary.

RIGHT *Make sure that birds have clean water and cannot drown. Wide, low buckets can get fouled with droppings, so large plastic 'duck drinkers' are a useful back-up, especially in hot weather.*

General Health and Care

Small numbers of birds kept in clean conditions should expect long healthy lives. Clean housing and litter, clean grass and water, and appropriate food and management will all prolong the life, and your enjoyment, of the birds.

Diseases are in the environment, however. Bacteria in the soil can affect susceptible individuals; wild birds carry bird diseases and parasites; and disease can travel on people and items connected with birds. Birds taken to shows, or new birds introduced to the premises, can also carry a slight risk of introduction of disease. It is therefore wise always to have an eye to biosecurity.

This is a term very much used since the foot and mouth epidemic of 2001 and avian influenza outbreaks in Europe since 2003. Strict biosecurity is observed on commercial premises where thousands of birds are reared in a short space of time. Disease can spread and intensify very rapidly in crowded conditions.

Where birds are kept in small numbers, 'biosecurity' just means keeping them in a sensible way: cleaning out dirty litter and composting it; minimizing access to food by wild birds; and restricting access of people, especially if they are bird keepers, in times of biosecurity alerts. This is sound advice given by government agricultural departments in charge of such matters to all keepers of animals and birds.

Observation each day

Spending some time with the birds each day is the best way to check that they are in good health. Birds which come to you for food, and are busy, talkative and bright-eyed, are healthy. If there is a change in their normal behaviour, or the colour of their beak and feet, then there is cause for concern. A slow walk, a limp, ruffled feathers and dull eyes are sure signs of problems. Such behaviour may be accompanied by weight loss, a drop in egg production, a dirty vent, green or loose droppings, lack of appetite or refusal to drink.

Handling the bird

If there is any concern about behaviour, first handle the bird to assess its condition before taking it to the vet. If a bird is really ill, it will not move away and you will be able to pick it up easily. However, it is surprising how much energy unfit birds can muster if they do not want to be caught.

The best way to catch the duck or goose is to drive it into its normal night-time accommodation, then corner it. Slip one hand under the body, using the other hand to hold a wing. Lift the bird from under the body, taking its weight on your hand, and tuck it under your arm with its head to the rear. This position restrains the wings and you can then use both hands to hold the bird by the thighs. Never catch waterfowl by the legs. You can grasp them by their strong neck to restrain them, but do not lift them this way either.

- **Assess the bird's weight,** especially the amount of muscle on the breast. Regular checks will tell you if a bird has become light, or is too heavy.
- **Examine the feel of the breast.** Lack of flesh and a sharp breast bone indicate that the bird is thin from lack of appetite and has an infection, or worms.
- **Examine the head and neck region especially** for parasites. Also check the eyes.
- **Check the vent for cleanliness.** A dirty vent is a sign of infection, or egg-binding in females.

- **If possible, take a sample of droppings** to the vet in a sealed, disposable container. If necessary, the vet can check for worms, worm eggs and coccidia.
- **Get a diagnosis from the vet** as quickly as possible. An early response to a problem can often cure it, whilst waiting only lowers any chance of survival.
- **Antibiotics can only be prescribed by a vet** and, of course, only combat bacterial diseases. A limited number of antibiotics and wormers are licensed for ducks and geese, but the vet can prescribe additional remedies for pet birds.

AILMENTS AND DISEASES

Aspergillosis

Symptoms: Lung congestion and laboured breathing.
Cause: Spores of mould from fungi growing in damp bedding (especially hay) or from mouldy food. Farming families may be familiar with this condition known as 'farmer's lung', for it affects people too.
Treatment: Fungal treatment may be available for birds, but this disease is best avoided by good management. Use wood shavings rather than chopped straw or hay for bedding. Keep incubators scrupulously clean. Dirty eggs can contaminate the incubator, and it is possible for *Aspergillus* to penetrate the egg and infect embryos. Young goslings may become infected during hatching from either dirty incubator equipment or dirty eggs.
Aflatoxin poisoning may show similar symptoms to aspergillosis. In this case, the moulds that grow on cereal grains and oilseeds produce toxins which are very damaging for ducks. Store food in dry, cool conditions; don't feed mouldy pellets or old bread.

Botulism

Symptoms: Loss of muscular control of legs, wings and neck, hence the term limberneck. Birds are unable to swallow.
Cause: Toxins produced by bacteria (Clostridia) in decaying animal and vegetable waste.
Treatment: Avoid problems by keeping ducks out of muddy/dirty areas and stagnant pools, especially in hot weather. The bacteria multiply rapidly in warmer temperatures in anaerobic conditions (where oxygen is excluded). Give affected birds fresh drinking water. If necessary, introduce water into the mouth and throat with a needleless syringe. A crop tube could be used with the advice of a vet. Add Epsom salts (magnesium sulphate) to the water. Recommended amounts vary from 1 tablespoon in one cup of water to 30g per 1.5l (1oz per 50 fl. oz/2½pints).

Bumblefoot
(Ulcerative pododermatitis)

Symptoms: The goose limps due to a ball of infected tissue under the foot.

Cause: Failure to resolve initial *Staphylococcus* infection with antibiotic treatment.
Treatment: Surgical removal of granular material by the vet. Ointments such as Leucagel may help.

Coccidiosis

Symptoms: Red blood in the droppings; most likely to be seen in early morning, when all food has passed through overnight. Check this by bedding birds on clean newspaper. Birds are thin because coccidia attack the lining of the gut, and nutrients from food are not absorbed.
Intestinal coccidiosis mostly affects young birds, which develop a tottering gait, debility, diarrhoea and bloodstained droppings. Birds may be ill for some time – weeks, not days.

Geese can also contract **renal coccidiosis** at 3–12 weeks. They become depressed, weak, walk slowly and drop their wings. Mortality can be high. Young birds are more susceptible than adults which have acquired immunity.
Cause: Ground dirty with droppings of birds which carry coccidia. Most likely to be a problem in summer.
Treatment: Coccidia do not respond to treatment by antibiotics. Give an anticoccidial

in the drinking water, or as a single drench (liquid down the throat), obtainable from the vet. The coccidiostat added to hen grower pellets is not a treatment for this condition.

Avoid problems by rearing young ducklings on clean ground, moving their protective coop to a new patch each day. Ensure that their water is clean and that they have a good diet of wheat and duck pellets. Coccidiosis is not common in ducks kept under these conditions. Geese are more likely to get it from grazing on dirty grass. Cider vinegar is often recommended as a deterrent.

Coccidiostats added to poultry food deter some of the parasites that cause coccidiosis, but evidence of their effectiveness with waterfowl is conflicting. One study found that waterfowl tolerated these medicated chicken rations; another case, recorded by breeders, showed that the additives almost certainly killed Call ducks, which may be more prone to intolerance. It certainly seems unlikely that the additives were specifically tested for use on these small ducks.

Co-ordination loss

This can happen in geese and the cause is not really known. Failure to use the legs can be due to leg infection, so check for temperature and swellings. Birds can also go 'off their legs' if they have been transported badly; they need time to recover. Older birds may be suffering from a damaged liver. In hot spells in summer, listeria has been cited.

Egg binding

Symptoms: A female will frequently sit down and look lethargic. She bobs her tail up and down as she strains to push the egg out. The vent is dirty, and some pink flesh may be pushed out, especially if the egg is low in the oviduct. If the egg is hard and has a good shell, it can be felt between the pubic bones.
Cause: Cold weather, cold rain, not enough calcium and magnesium in the diet. These minerals are needed for good shell formation.

Poor eggs may have only a membrane and no shell, or a rough shell. Eggs with a strong, smooth shell are easier to lay than soft-shelled or rough-shelled eggs. Calcium is also needed for muscle contractions to push the eggs down and out of the oviduct.
Treatment: Long-term – provide a good diet of layer pellets for females, with the correct ratio of calcium and magnesium. Also supply mixed poultry grit. Lime the ground with slaked lime or calcified seaweed if you live in a high rainfall area, especially if the soil is acid and lime does not occur naturally.

Short term – bring an affected female into a warm environment of 24–28°C (75–82°F). Give extra calcium in the form of Calcivet from The Bird Care Company. This is administered orally and can quickly bump up the calcium supply. Note that this is a short-term treatment for calcium deficiency and is not a remedy for frequent soft-shelled eggs. Calcium is also easily found in indigestion tablets: read the ingredients. If this fails, a calcium injection, given by a vet, may help.

Tame birds can be helped by floating them in a bath of warm water and pushing the egg (felt under the abdomen) outwards at the same time as the bird pushes.

In a bad case where the egg will not pass out, but can be seen at the vent, the exposed shell can be broken with forceps, the contents extracted by syringe, and the shell gently pulled out. The bird is more likely to suffer from an infection if the eggshell breaks, so this is a last resort at the vet's and must be followed up with a course of antibiotics.

In a bad case of a bird straining to pass an egg, the oviduct will prolapse i.e. turn outwards. This reduces the bird's chance of survival. If the egg is successfully passed without prolapse, subsequent eggs may be passed normally and the bird never experience problems again. Sometimes egg problems occur high in the oviduct and nothing can be done to save the bird, which is best put down. Egg problems are more common in Call ducks than in any other breed.

Dropped tongue

Symptoms: The orange skin of the lower mandible drops down, so that the bird appears to have a second 'beak'. There is difficulty in feeding, especially grazing. Head shaking and loss of weight also occur.
Cause: In heavier breeds of geese, especially those with a dewlap, food and grit can get stuck under the tongue. This causes the soft centre of the lower mandible to drop; the tongue also drops, exacerbating the problem. Feeding mash will make matters worse.
Treatment: This condition is rare, but any bird showing this problem should first have their mouth cleaned. If the problem does not resolve, a vet can stitch the loose skin. Suturing and surgical removal are better, reducing any chance of infection. The operation is completely successful.

Enteritis

Inflammation and bleeding in the gut can be produced by bacteria or a virus.

Duck viral enteritis (DVE) can occur in ducks, geese or swans, but is rare in all three, particularly geese. It will, however, kill most affected birds.
Symptoms: Birds are listless and suffer from pinkish droppings. The bill turns blue (cyanosis).
Cause: A herpes virus carried by wild mallard; they should be discouraged.
Treatment: Prompt treatment with a vaccine obtainable from your vet is the only solution. This will then protect the remaining birds. Birds that have recovered from DVE are immune to re-infection.

Bacterial enteritis
Symptoms: Lethargy, loss of appetite; birds still drink. Pinkish droppings.
Cause: Probably carried by wild birds. Affects ducklings; older birds seem to be immune.
Treatment: Soluble antibiotic powder from the

vet, in the drinking water, but you must catch this early. No other water should be available. Move the birds onto clean ground. Avoid problems by rearing ducklings on clean ground.

Lameness

Symptoms: Limping; hot leg, swollen ankle or swollen hock.
Cause: Bacterial infection e.g. *Staphylococcus*.
Treatment: Course of antibiotic injections prescribed by a vet.

Malignant growths

Geese, more often than ducks it seems, suffer from tumours. These can occur on the wings, on the head and under the body, where they can be felt in the soft parts. Sometimes the tumour is internal and the bird is abnormally heavy and ill. Birds in this condition should be put down humanely.

Tumours are, nevertheless, uncommon. Care must be taken not to confuse a sinus infection on the face with a tumour.

Mites and lice

Symptoms: Birds scratch a lot. Northern mite lives on the bird and sucks its blood. This is different from the red mite which lives in the chicken shed. Lice, which are insects, also live on the birds. They are found on bits of feather, and are grey in colour instead of red. Check the head and the vent region, especially in old or sitting birds.
Cause: Northern mite is caught from other birds at bird shows, and from introduced, new birds. It is possible that it can also be caught from wild birds.
Treatment: Use a pesticide. Johnson's Flea Powder is licensed for cats, dogs and cage birds. Ivermectin (pour-on) for controlling internal and external parasites is now available from vets in 10 ml dropper bottles for pigeons. Ask your vet for advice on both these products.

Barrier red mite powder is an organic alternative, and can be bought from agricultural stores. It contains only plant oils as active ingredients, and maize as a carrier.

Note that pesticides have to be used at least twice to eliminate parasites because the eggs are not affected. Thus where the parasite's lifecycle (from egg to reproducing adult) is 7–10 days, as in the case of the red mite, the birds should be treated 2–3 times at 8–10 day intervals. Organic, herbal-based pesticides are effective as long as they are in date and used regularly.

Lead poisoning

Symptoms: Lack of co-ordination, loss of weight.
Cause: Lead shot from cartridges or air-gun pellets, picked up by the birds as 'grit'.
Avoidance: Make sure the source of lead cannot be accessed. Provide grit for the birds so that they do not pick up bits of lead instead.

Pasteurella
(Fowl cholera; pasteurellosis)

Symptoms: Loss of appetite, increased thirst, watery then green droppings. Loss of co-ordination.
Cause: Bacteria in the environment affect vulnerable birds.
Treatment: Prompt treatment with intramuscular antibiotic injections or soluble drugs from the vet may save larger birds. Smaller birds usually succumb. Eliminate carriers, such as rats.

Prolapse

Symptoms: Males – the penis is dropped externally from the body. Much more common in drakes than in ganders. Females – the lower part of the oviduct protrudes.
Cause: In males, too much mating with the females. In females, difficulty in passing eggs. See **Egg binding,** opposite.

Treatment: The prolapse can be gently pushed back inside if an egg has been laid, but quite often the problem recurs and the bird is best put down.

Respiratory problems

Symptoms: The bird sits hunched up and bobs its tail up and down to assist in breathing.
Cause: Bacterial infection, especially in spells of intensely wet weather. The symptoms of Aspergillosis are similar, but the fungal disease will not respond to antibiotic treatment.
Treatment: A long course of antibiotic from the vet. Birds' lungs are complicated, because of adaptations for flight. So an infection is difficult to resolve.

Sinus problems

Symptoms: Weeping nostrils and puffed-up cheeks.
Cause: Bacteria in the environment infect the sinuses. More prevalent in Calls than in other breeds of ducks.
Treatment: Appropriate antibiotic injection obtainable from the vet; Baytril is often prescribed, but another more effective antibiotic may be available. Treatment should be immediate. If left, the cheeks harden and the bird cannot be cured.

Slipped wing

Symptoms: Young birds' primary feathers turn outwards. They may also just drop.
Cause: The young birds are fed a diet too high in protein and grow too fast. The blood in the quills is too heavy for the wings to support. This phenomenon does not occur in wild birds.
Avoidance: Avoid this problem by feeding growing birds a lower protein diet while they develop the primary feathers, between five and eight weeks. Change the breeding stock to stop this problem developing again.

Wet feather

A bird's feathers look dirty, untidy and wet. See page 19.

Worms

Stock should be wormed routinely twice a year, and on any other occasion which necessitates it e.g. if a bird seems ill or is coughing. Ducks seem to suffer less from worms than geese, but any bird which is underweight should be treated. Signs of worm infestation include loss of weight (lack of breast muscle and a sharp sternum) and bloodstaining around the vent. If a bird is wormed and placed on newspaper (in a small pen) for the night, round worms may be seen in the droppings. Collecting a sample of droppings will enable the vet to identify the species of worm eggs present under the microscope.

Worms which affect waterfowl come in a variety of forms.
- Gizzard worm (*Amidostomum*) is more likely to be lethal in geese.
- Gapeworms (*Syngamus*; *Cyathostoma* in geese) make birds cough and, in extreme cases, will asphyxiate them.
- Round worms (*Ascarides*) live in the gut. Occasionally seen in droppings.
- Caecal worm (*Heterakis*) inhabit the caeca (two blind-ending extensions from the gut).
- Tapeworm and fluke.

Most of these worms use earthworms or insects as a host, and wild birds are carriers. So, however clean the environment, there are always some parasites present. The higher the density of stocking and the greater the length of time over which the land has been used, the greater the importance of regular worming and good ground management. Liming ground, exposing it to sunlight and rotating stock all help.

RIGHT: *A pair of Blue Swedish ducks enjoying life in the garden.*

Treatment

The preferred wormer for birds is Flubenvet, a white powder which can be obtained from your vet. It comes in a 240g (8½oz) tub and usually has a very long use-by date. It is licensed for use on birds and, at the correct dosage, kills all the parasites listed above. Ducks are not mentioned on the label, so check the dosage with a vet. The dosage for geese and chickens is 120g (about 4oz) on 100kg (220lb) of food. This works out at 1.2g per kilo or about $^1/_{10}$ oz per 5lb – easier to measure at one level teaspoonful (3.6g or ⅛oz) per 3kg (6½lb). Check the weight of a teaspoonful on digital kitchen scales. The powder adheres well to pellets (better than to wheat), so feed mostly pellets over the worming period. Mix the Flubenvet into the food with a tablespoon – don't use your hand, because the powder sticks to your skin. The disadvantage of Flubenvet is that you have to feed it for a week for it to be effective.

Birds which are really ill, and not eating, cannot be dosed in this way. Another product, licensed for cage birds and obtainable from a vet, that is delivered as a single drench (liquid down the throat), is more suitable. This is also a better vermifuge for geese, whose diet is mainly grass, not pellets.

The number of worm species that Ivermectin (see Mites, page 113) kills is also more limited than Flubenvet – it doesn't work with tapeworms or fluke. But it is doubly useful in that it systemically kills external parasites, such as northern mite.

This is not an exhaustive list of ailments and diseases. It excludes, for example, avian influenza, TB, crop binding, salmonella, Newcastle disease etc, many of which will not occur in small flocks of pet birds. Ducks and geese suffer from fewer diseases than chickens. Waterfowl do not require vaccination for Mareks, Gumboro and infectious bronchitis.

Note that if medicines are given to ducks and geese the appropriate withdrawal time (usually a week) should be followed before eggs are consumed.

Using Eggs and Feathers

Waterfowl can open up new interests and hobbies – such as herding the Quack Pack, which can be seen at summer shows. Feathers and down are by-products of table birds, and goose quills and eggs can be put to decorative as well as practical use.

Calligraphy is an ancient art, and decorated goose eggs probably inspired Fabergé. Simply paint the eggs, or use stickers and découpage. Or be more adventurous and check out the Egg Crafters' Guild of Great Britain for advice on how to make intricate designs.

A final word. Duck eggs really are the best for baking. Do use these recipes to find out.

Painted and Decorated Eggs

Eggs come in a huge variety of sizes, depending on the species of bird. Traditional Easter eggs seem to have been painted hen eggs and, where egg-rolling games were played, the eggs were just hard-boiled and painted.

When eggs are destined to be kept, they are first blown of their raw contents so that the shell alone is preserved.

Exquisite eggs are produced by the Polish method of either covering the shell with a layer of molten wax and then etching the design or drawing with wax on the bare egg. Colour is added by submerging the egg in dye. A similar effect can be achieved more easily by using layers of acrylic paint. Découpage – sticking on small paper cut-out shapes – is also an effective way of decorating eggs.

Blowing goose eggs

Duck eggs are not very suitable for blowing because the shells are not strong enough. Goose eggs are much better, but need special treatment because of the thickness of the shell and the stiffer, more gelatinous albumen, compared with hen eggs. However tempting they may be, the very largest double-yolked goose eggs may not work well. The shell is often not strong enough.

1. First puncture the shell at the pointed end with a very sharp compass point or needle, or a fine bit on an electric drill. Start with a small hole and aim to enlarge it later.
2. Then puncture the egg at the bulbous end. The hole will need to be 4mm (⅙in) in diameter to get the contents out; it needs to be larger at the exit than the end where you blow in.
3. Insert a knitting needle into this larger hole to break up the contents, and shake the egg. If the membranes are intact, blowing will be impossible.
4. Using the smaller hole, blow out the contents into a bowl. Be careful of the pressure on your ears.
5. Wash the shell out thoroughly, using dilute bleach or Milton. Dry the egg out completely afterwards.
6. If the holes are too large, they can be filled later with decorative materials.

If you are blowing a large number of eggs, use a pump to evacuate the contents.

Design

The much larger eggs of goose, emu and ostrich have a thicker and stronger shell and so can be carved into intricate designs. They then have beads and fake jewels attached to them. Fabergé eggs produced for the Russian court, and made of precious gemstones and gold, are in this style.

Before cutting the egg shell, design the desired effect, then trace this lightly in pencil on the shell. Suitable cutting implements are scalpels and fret blades which can enter the holes already drilled. You can now buy small electrical cutters with a variety of attachments for drilling and cutting. Embellishments for the eggs can be found in craft and specialist sewing shops and also in catalogues online. Painted and decorated eggs can be exhibited in competitions at bird shows; goose eggs are the most popular type for this purpose.

Feathers and Down

Feathers, down and quills from geese and ducks were an essential part of everyday life before the inventions of the oil-refining industry. These natural products are still used today. Some are collected in a sustainable manner; others are produced in a more dubious way in welfare terms.

Feathers and down from ducks and geese have been much prized for bedding in the past. Poor people had to stuff their mattresses with straw, chaff or bracken, but those who could afford it chose the feather bed. Whilst the sprung mattress has replaced the feather one, the best duvets and pillows are not polyester – they are still made from eider (a wild sea duck) or goose down.

Down from the eider is now harvested under licence in Iceland from sustainable nesting areas. Goose down is a by-product of the goose-rearing industry in Eastern Europe. As well as being plucked when they were killed, the unfortunate birds used to be 'live plucked' as much as three times a year to maximize the amount of best quality feather and down that could be produced. According to an FAO report, this practice persisted until the 1990s in Poland.

For home-preserving of down, baking in the oven has been recommended, perhaps similar to this account from 1902: 'The plan most generally pursued by farmers' wives . . . is to lay the feathers on paper on the floor of a spare room to dry, and then beat them lightly to get out the dust or dirt, after which the quill parts are carefully cut off, and then they are put into bags and placed in the oven after the baking, while it is yet slightly warm, and left there, when, after a few weeks, the bags are hung in rows on beams in an airy room so as to get thoroughly free from all moisture, and when that is so, are put into bed, pillow or cushion "ticks" for use.'

Flights

When the longbow was an essential part of warfare, at the time of the battles of Crécy (1346) and Agincourt (1415), the goose played a crucial role. Archers would select flight feathers to 'fletch' their arrows. These flights could have come from the farmyard goose or, more likely, the greylag.

Feathers in calligraphy

In the past, reeds were used for inscribing clay tablets and as a pen for writing on papyrus. Some of the Dead Sea Scrolls are said to be written in quill pen, and the technique was introduced to Europe by around 700 AD. By medieval times the goose quill was the preferred method for writing on parchment in ink, though crow and swan feathers were also used. The best quality feathers were pulled from living birds; feathers and quills are not as strong after a year's wear and tear and moulting. The favoured feathers for writing are the outer flights, which are more asymmetrical and also larger than the inner ones. The wing feathers curve inwards, so right-handed people might prefer flights taken from the left wing, which curve away from the hand.

Preparing a quill pen

Goose feathers – the flights – can be bought at craft stores or online. Buy several because they are not all the same quality; practise cutting and preparing the poorer quality ones first. The ones with a wider shaft have a better reservoir for ink. The shape of the feathers will also vary

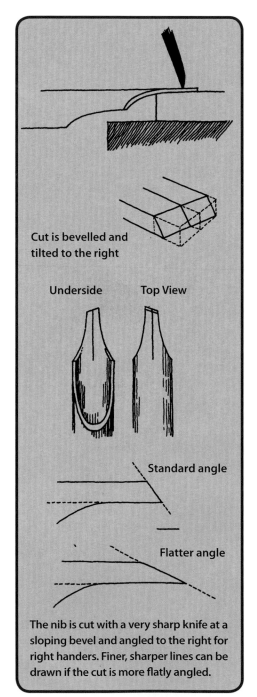

Cut is bevelled and tilted to the right

Underside   **Top View**

Standard angle

Flatter angle

The nib is cut with a very sharp knife at a sloping bevel and angled to the right for right handers. Finer, sharper lines can be drawn if the cut is more flatly angled.

depending on which part of the wing they came from. If you keep geese, you can select your own quills each year from their moulted feathers.

Quill pens can carry the whole plume but in the past the barbs were often partially or even fully stripped away. Before cutting the nib, decide how the quill should look. It is best to strip away any fluff near the nib end to keep the ink and writing cleaner.

The end of the quill is almost closed, so it needs to be cut to allow ink into the reservoir, and also to make a nib. Place the quill on a kitchen board and make a sweeping, oblique cut on the underside of the quill, then cut the nib longitudinally through the centre to ease the flow of ink. The tip itself can be cut across to create a square-cut nib.

A knife has to be sharp to cut a goose quill and the result is almost as good when the quill is cut obliquely with very strong, sharp scissors.

The feather shaft becomes harder if it is baked. Put a can of sand into the oven at 175–190°C (350–375°F) and leave it until the sand is hot all the way through. Take the can out of the oven and stick the quills, point down, into the sand as far as they will go. Leave them there until the sand cools. The feather tips will now be hardened. This is sometimes advocated before cutting the nib, but I find the quill cuts more easily untreated.

The hand-cut goose quill is still considered the best calligraphy tool, providing a sharp stroke and flexibility unmatched in steel pens. Practise on thicker paper such as water-colour, which will absorb the ink or black paint easily. Then move on to copier paper which is smoother, like parchment. Only use expensive materials when you are well practised.

Aa Bb Cc Dd Ee Ff Gg Hh Ii Jj Kk Ll Mm Nn Oo Pp Qq Rr Ss Tt Uu Vv Ww Xx Yy Zz

Goose and Duck Egg Recipes

Duck eggs are in high demand as speciality food items. Like all eggs they need to be collected from clean nests, used within two weeks and cooked properly. They are delicious and better for purpose than hen eggs.

The average duck egg from an Indian Runner, around 70g (2½oz), is about the same size as a large hen egg. Of course, eggs vary a great deal in size between the breeds, so duck eggs should be weighed to check the number needed for each recipe.

Duck and goose eggs contain more protein and fat than hen eggs, and less water. They taste creamier and make far superior cakes. It is said that the whites do not whisk up as easily as hen eggs. However, when whisked for these cake recipes, they remove the necessity for raising agents and produce a better-tasting product with no bitterness. That they are the best for the job is reflected in their price: duck eggs sell at a premium.

They are also much in demand by the Chinese community where they are used for salt pickling in the traditional South-East Asian style.

Orange and almond cake

Ingredients for cake
6 duck eggs, separated, total weight 400g (4oz), or separated goose eggs of similar total weight
240g (9oz) sugar
200g (7oz) ground almonds
50g (2oz) semolina
grated zest of an unwaxed lemon or orange (ideally Seville orange, in season)
flaked almonds, to decorate

Ingredients for syrup
juice of 4–6 oranges
juice of one lemon
sugar to taste

Heat the oven to 180°C (350°F). Line a cake tin with greaseproof paper. Combine the egg yolks with the sugar. Beat well with a wooden spoon, then add the lemon/orange zest, almonds and semolina. In a separate bowl, whisk the egg whites until stiff. Fold into the yellow sugar and egg mixture with a metal spoon until well mixed. Pour into the prepared case, sprinkle the flaked almonds on top and bake in the middle of the oven for about an hour. It should be firm to the touch.

Put all of the ingredients for the syrup into a pan and bring to the boil. Simmer for a few minutes to thicken, and then pour over the warm cake.

Coffee cake

This is the standard recipe for Victoria sandwich, flavoured with coffee essence. What makes it so special is quality organic, wholemeal flour such as Bacheldre Mill, and the whisked whites of the duck eggs.

Ingredients for cake

180g (6oz) sugar
180g (6oz) margarine
3 duck eggs, separated (around 70g/2½oz each)
180g (6oz) wholemeal organic flour
coffee essence: 2 tsp coffee granules dissolved in 3 tsp hot water
cherries and/or walnut halves for decoration

Ingredients for icing

icing sugar
knob of butter or margarine to make a creamy texture
coffee essence: 2 tsp coffee granules dissolved in 3 tsp hot water
small amount of additional water to mix

Heat the oven to 180°C (350°F). Line a cake tin that is about 20cm (8in) in diameter with greaseproof paper so the cake can be removed easily.

Using a wooden spoon, mix and beat the sugar and margarine until pale and creamy. Beat in the egg yolks one at a time until well combined. (It may help to crack the eggs into a small basin and remove the yolk with a soup spoon.)

Whisk the separated egg whites, then using a metal tablespoon, gently fold in the whisked whites, taking care to keep air in the mixture. Fold in the sifted flour, and finally mix in the coffee essence, using as little water as possible to dissolve the coffee granules.

Pour the mixture into the prepared case and bake in the middle of the oven for about 40 minutes. It should be firm to the touch.

Remove the cake from the tin when it is cool. Mix the icing ingredients: the butter or margarine stops the icing from being runny. Cut the cake in half and use some of the icing mixture to sandwich it back together, and spread the rest on the top. Decorate with cherries and/or walnut halves.

Goose egg custard – *the* crème caramel

Goose eggs make the best crème caramel ever. Geese kept on grass (as they should be) produce eggs with a naturally bright orange yolk which gives a very good colour to the custard. The eggs also give an exceptionally creamy texture.

Ingredients

2–3 goose eggs weighing around 140–210g (5–7½oz)

600ml (1pt) milk (semi-skimmed or full fat)

1 tablespoon sugar, plus 150g (5oz) for the caramel

Make the caramel first. Pre-heat a Pyrex dish large enough to take the ingredients. Dissolve the sugar in about 75ml (3fl. oz) of water in a saucepan and then boil vigorously until the liquid starts to brown. Do not stir with a spoon. Cautiously continue to heat the syrup until it reaches the desired colour. If overdone, it will burn. Stand the heated dish on an insulated surface such as a dry tea towel folded on a board, then quickly pour the very hot liquid into it. The caramel will set almost instantly. It can be made in advance and used when necessary.

Pre-heat the oven to 120°C (250°F). Make the custard by whisking the eggs with the milk and sugar in a saucepan. Gently heat the mixture to around 40°C (104°F, a little above blood heat), stirring so the custard does not cook on the bottom of the pan. Then pour the mixture into the glass dish and sit the dish in a water bath in the oven. The pre-heating of the custard and the use of a waterbath speed up the cooking, but slow down excessive heating. With the oven set at around 170°C (340°F), the cooking can take one hour. The custard must be set but not boil; it ruins the texture, so 120°C (250°F) for two hours is better. (Note that the cooking time will be less for smaller quantites.) Check if the custard is set by shaking the dish very gently.

The dish can be cooled, then chilled and will keep for several days.

Pickled eggs

This is a good way of preserving surplus eggs, and the product (in a decorated jar) can be used for a present. Pickled eggs are easy to produce and are a great accompaniment to salads, cold meat, fish and chips, or sandwiches.

Ingredients
Eggs
Vinegar: brown malt or clear distilled
black peppercorns, rosemary, whole chillies, garlic cloves, lemon rind

The quantity depends upon the number of eggs you're pickling and the size of your storage jar. As a general rule of thumb, you should have enough vinegar to half fill the storage jar when empty.

Hard-boil the eggs for ten minutes. Drain and cool in cold water. Shell when cool. It is better to use eggs which are not very freshly laid: the shell sticks to the albumen. Eggs that are two weeks old will shell cleanly.

Pack the eggs into a clean glass jar, preferably one with a rubber seal.

Meanwhile prepare enough vinegar to cover the eggs in the jar. You can use any vinegar but standard malt vinegar tinges the eggs brown; they look better in clear vinegar. Put the vinegar in a saucepan with a sprinkling of peppercorns, chillies and a bay leaf. Bring to the boil and simmer for a few minutes. Pour the liquid over the eggs while it is still warm: this creates a good seal for preserving.

The usual recommendation is that the eggs are left for a month before eating, but they can be eaten straight away. Store the bottle in the fridge if there is no seal. In a sealed jar the eggs will keep for six months.

You can try making up your own pickling liquid combination according to your own taste – beetroot and ginger are very good additions to experiment with.

Useful Address and Links

Feed Suppliers

Allen & Page Smallholder Feeds
Natural feed without nature-identical ingredients or flavours (that is, artificially manufactured to mimic nature), artificial growth promoters or artificial yolk pigmenters.
Tel: 01362 822900
www.smallholderfeed.co.uk

BOCM-Pauls Ltd–Marsdens Feeds
PO Box 2, Olympia Mills, Barlby Road, Selby YO8 5AF
Tel: 01757 244000 or 08457 165103
and BOCM PAULS LTD, Colomendy Industrial Estate, Denbigh, LL16 5TA
Tel: 0845 0250 444
enquiries@bocmpauls.co.uk
http://www.bocmpauls.co.uk/bocmpauls/marsdens/homepage.jhtml

Charnwood Milling Company Ltd,
Specialist diets for all waterbirds.
Framlingham, Suffolk IP13 9PT
Tel: 01728 622300 www.charnwood-milling.co.uk

Marriage's Feeds
A well-known independent family milling business established in Chelmsford in 1824, where the family had already been farming and milling for at least 100 years.
Tel: 01245 612000 www.marriagefeeds.co.uk

Other Products

Ascott Smallholding Supplies
A family-run firm offering a wide range of equipment for smallholders and poultry keepers. Poultry equipment includes incubators and brooders, suitable for all garden poultry including hens, ducks and waterfowl.
http://www.ascott.biz

Barrier Biotech
www.barrier-biotech.com

Brinsea Incubators Ltd
www.brinsea.co.uk

Forsham Cottage Arks (Middle England)
Haws Hill Farm, Sutton, Tenbury Wells WR15 8RJ
Tel: 01885 410300
enquiries@forshammiddleengland.co.uk
http://www.forshammiddleengland.co.uk

Interhatch
Incubator specialists – sales, service, repairs and spares for most makes of incubators. Waterfowl equipment.
Whittington Way, Old Whittington, Chesterfield S41 9AG
Tel: 01246 264646

The Bird Care Company
www.birdcareco.com

Magazines

Country Smallholding Magazine
http://www.countrysmallholding.com

Fancy Fowl
The longest serving magazine in the UK for poultry and waterfowl information for showing and pleasure
http://www.fancyfowl.com

Poultry Press
PO Box 542 Connersville, Indiana 47331-0542
Tel: 1-765-827-0932 Fax: 1-765-827-4186
http://www.poultrypress.com

South East Asian Poultry
Editor: Megg Miller. Contact Poultry Information Publishers, PO Box 438, Seymour, Victoria 3661, Australia
Tel: (03) 5792 4000 Fax: (03) 5792 4222

Smallholder Magazine
Editor Liz Wright.
http://www.smallholder.co.uk

Picture Credits

Main photography by Chris and Mike Ashton.

Pages 4–5 © Ardea.com; 21 © Ardea.com; 23 © Alamy; 24–25 © NHPA photos; 26–27 © NHPA photos; 31 © Ardea.com; 33 © Ardea.com; 34–35 © Corbis. All rights reserved; 41 © Alamy; 45tl © NHPA photos; 50 © Ardea.com; 72 © Alamy; 90–91 © Marie O'Hara; 96 © NHPA photos; 103 © Ardea.com; 122 © Alamy; 123 © Alamy; 124 © iStockphoto.com; 125 © Corbis. All rights reserved.

Special thanks to fellow waterfowl enthusiasts who have supplied photographs:

59 Buff Back geese – Graham Barnard
63 American Buff, 65 West of England , 71 Aylesbury, 76 Cayuga, 87 Silver Bantam Duck – Simon James
74 Pekin – Bart Poulmans
78 Rouen Clair – Mike Sumner
98 Brinsea Products

93 adapted from Proctor N S & Lynch, P.J. 1983

Bibliography

The following authorities are quoted in the text and the list below gives suggested further reading for those interested in pursuing the subject of ducks and geese in more depth.

William Ellis, 18th-century Herefordshire farmer and writer on country matters. *The Country Housewife's Family Companion* provides a remarkable insight into the food enjoyed by fairly humble farming families in southern England at the time. http://www.historicfood.com/Wiggs%20recipe.htm

Konrad Zacharias Lorenz (1903–89), Austrian zoologist, animal psychologist, ornithologist and Nobel Prize winner: one of the founders of modern ethology. He studied instinctive behaviour in animals, especially in greylag geese.

Gervase Markham, prolific writer on country affairs in the 17th century.

Harrison Weir (1824–1906), natural history artist and author on poultry and waterfowl.

Francis Willughby (1635–72), English ornithologist. At Cambridge he was taught by the naturalist John Ray. Ray published Willughby's *Ornithologia libri tres* in 1676, with an English edition two years later.

Lewis Wright, Victorian author on poultry and waterfowl

Ambrose, Alison — *The Aylesbury Duck* (Buckinghamshire County Museum, 1991, 2nd revised edition)

Ashton, C & Ashton, M — *The Indian Runner Duck – A Historical Guide* (The Feathered World, Winckley Press, Preston, 2002)

Bowie, S H U — 'Shetland's native farm animals, Part 3 – Shetland geese' The Ark (Rare Breeds Survival Trust, April 1989)

British Waterfowl Association — *British Waterfowl Standards* (2008) www.waterfowl.org.uk

Darwin, Charles — *The Variation of Animals and Plants under Domestication*, Vol. 1 (1868)

Delacour, J. — *The Waterfowl of the World* (Country Life, London, 1964)

Ellis, William — *The Country Housewife's Family Companion* (London, 1750)

Freethy, Ron — *How Birds Work* (Blandford Press, Poole, 1982)

Ives, Paul — *Domestic Geese and Ducks* (Orange Judd, New York, 1947)

Lorenz, Konrad — *The Year of the Greylag Goose* (Eyre Methuen, London, 1979)

Markham, Gervase — *The English Husbandman* (1613)

Moubray, B. — *A Practical Treatise on Breeding, Rearing, and fattening all kinds of Domestic Poultry, Pheasants, Pigeons and Rabbits* (1815)

Proctor N S & Lynch, P J — *Manual of Ornithology* (Yale University Press, 1993).

Weir, Harrison — *Our Poultry* (Hutchinson & Co., London, 1902)

Willughby, Francis — *Ornithology* (1678)

Wright, Lewis — *The Illustrated Book of Poultry* (Cassell, Peter & Galpin, 1873), 1st Edition

Glossary

Autosexing Sexual dichromatism in the colour; males are a different colour or pattern from the females.

Bantam ducks Smaller ducks, weight below 1.4kg (3lb); down to 140g (1lb) in the Call duck.

Breed Variety with genetically stable phenotype; such as a pair of birds breeding true to type and colour.

Clipping The primary feathers of one wing are cut midway along their length – this will unbalance a bird which attempts flight.

Dewlap Fold or flap of loose skin below the lower mandible (bill).

Down Small feathers beneath the contour feathers with high insulation value; they lack interlocking barbiceli.

Gizzard Large muscular stomach where food is ground down with the aid of small stones (grit).

Heavy ducks UK Classification: Ducks over 3.2kg (7lb) (female), 3.6kg (8lb) (male).

Incubator fluff Yellow, fluffy material shed at hatching – the protein coating shed from down as it dries out and fluffs up.

Keel Deep, pendant fold of skin and tissue suspended from the breast bone and continuing under the body.

Light ducks UK Classification: Ducks under 3.2kg (7lb) (female), 3.4kg (7½lb) (male); above Bantam weight.

Medullary bone Bone formation present in the marrow spaces of the long bones of birds, which serve as a readily mobilised source of calcium for shell formation.

Moult Annual shedding of feathers in geese; twice-yearly loss of feathers in ducks.

Primary feathers Outer flight feathers, beyond the carpal joint or wrist.

Precocial young The young are relatively mature and mobile from the day of hatching. The opposite developmental strategy is called *altricial*, where the young are born helpless. Precocial species are normally nidifugous: they leave the nest shortly after birth.

Secondary feathers Inner flight feathers attached to the radius/ulna region of the wing; the speculum in ducks.

Species A group of organisms capable of interbreeding and producing fertile offspring.

Strain Examples of a breed kept over several generations by an individual breeder.

Type General shape or form.

Vent Orifice through which droppings pass (anus).

Wild-colour The colour of the wild species – mallard, greylag, swan goose – without colour mutations.

Index